Crop Production Technology-I
(*Kharif* Crops)

NIPA GENX ELECTRONIC RESOURCES & SOLUTIONS P. LTD.
New Delhi-110 034

About the Authors

Dr. M. Mohamed Amanullah, Professor of Agronomy (Retired) completed B.Sc. (Ag.) at Agricultural College and Research Institute, Madurai, and M.Sc. (Ag.) and Ph.D. in Agronomy at Agricultural College and Research Institute, Tamil Nadu Agricultural University, Coimbatore. He worked at Veterinary College and Research Institute, Namakkal, Tamil Nadu for 13 years and at TNAU for 20 years. During this period, apart from teaching undergraduate and postgraduate students, he also served as, UG and Research Coordinator. He has conducted 10 training programmes under Centre of Advanced Faculty Training of the ICAR as Course Coordinator. His teaching experience spans for 29 years to UG, PG and Ph.D. degree programmes. He has handled more than fifteen different courses for undergraduate and postgraduate students. He has guided eight M.Sc. (Ag) students and 5 Ph.D. students and one PDF student. He has published ten books with ISBN, 7 books without ISBN, published more than 240 research articles both in international and national journals apart from many popular articles. He has completed for external funded projects and many university research projects.

Dr. K. Rajendran, completed his U.G, P.G and Ph.D. in Tamil Nadu Agricultural University, Coimbatore. He has worked in TNAU as professor for more than 33 years in different positions covering Krishi Vigyan Kendra, Research Stations involving in teaching, research and extension. He was awarded ICAR Junior Research Fellowship for his P.G studies. He also visited International Rice Research Institute Manila Philippines under INSURF Fellowship. He has published more than 35 research articles in reputed peer reviewed journals both national as well as international. He has participated and presented research papers in seven international conferences and 10 national seminars covering bt cotton and hybrid rice. He has guided 7 M.Sc. (Agronomy) and 5 Ph.D. students. Currently he is working as Professor & Head of the Division of Agronomy at Karunya University, Coimbatore.

Dr. S. Marimuthu, completed B.Sc (Agriculture) at Agriculture College and Research Institute, Madurai and PG, Ph. D., degree programmes at Tamil Nadu Agricultural University, Coimbatore. He has published 20 research articles in reputed International and National Journals. He has participated in 5 International and 10 Natioanl conferences and won several awards and also he had guided 3 projects in UG students of B. Sc (Agri.) programme, having 3 years of research and 5 years of teaching experience. He is currently working as Assistant Professor (Agronomy) in SRM College of Agricultural Sciences, SRM Institute of Science and Technology, Tamil Nadu, India.

Crop Production Technology-I
(*Kharif* Crops)

M. Mohamed Amanullah
(Retd.) Professor (Agronomy)
Tamil Nadu Agricultural University
Tamil Nadu

K. Rajendran
(Retd.) Professor (Agronomy)
Tamil Nadu Agricultural University
Tamil Nadu

S. Marimuthu
Assistant Professor (Agronomy)
Faculty of Agricultural Sciences
SRM Institute of Science and Technology
Tamil Nadu

NIPA GENX ELECTRONIC RESOURCES & SOLUTIONS P. LTD.
New Delhi-110 034

NIPA GENX ELECTRONIC
RESOURCES & SOLUTIONS P. LTD.

101,103, Vikas Surya Plaza, CU Block
L.S.C.Market, Pitam Pura, New Delhi-110 034
Ph : +91 11 27341616, 27341717, 27341718
E-mail:newindiapublishingagency@gmail.com
www: www.nipabooks.com

For customer assistance, please contact
Phone: + 91-11-27 34 17 17 Fax: + 91-11- 27 34 16 16
E-Mail: feedbacks@nipabooks.com

ISBN: 978-93-94490-45-1

Composed and Designed by NIPA.

TAMIL NADU AGRICULTURAL UNIVERSITY

COIMBATORE - 641 003, TAMIL NADU, INDIA

Dr. K.S. SUBRAMANIAN, Ph.D.(Canada)

Director of Research

FOREWORD

I am pleased and deeply delighted to provide a foreword to a text book on **"Crop Production Technologies-I (*Kharif Crops)*"** which is designed and articulated as per the ICAR syllabus to fulfill the curriculum requirement of agricultural students while serving as a reference guide for the scientists and other users. The agronomy has taken a series of developments and modifications in the context of digital agriculture encompassing drip fertigation, precison agriculture and complete mechanization. These areas are of prime importance and the update is required to serve as a text book for the students and the knowledge base for scientists.

This book is orchestrated to meet the syllabus of stipulated by the ICAR V Deans' committee's recommendation and has a wide coverage of topics including origin, distribution, area, production, economic importance, soil, climate, season, seed rate, spacing, nutrient management, weed management, water management, cropping system and yield. This book is written in a simple language enabling the readers who can easily understand the concepts.

I congratulate the authors of this book Professor **Dr. M. Mohamed Amanullah** and **Dr. K. Rajendran** who have vast knowledge and a wide experience in undergraduate and post-graduate teaching and **Dr. S. Marimuthu,** Assistant Professor for the pain taking efforts in bringing out this text book. The model Question paper given at the end of each chapter will be useful for the students for preparing for examination.

12/5/2022

K.S. SUBRAMANIAN

Tel (D): +91 422 - 661 1447, 547, Tel (0): +91 422 - 661 1526, 527, 546, Mobile: (P) +91 98940 65449 / (0) +91 94890 33000

E-mail: drres@tnau.ac.in, kss@tnau.ac.in | Website: www.tnau.ac.in.www.agritech.tnau.ac.in | Fax: 0422-2436325

Innovate to Invigorate

Preface

Agriculture is the most important sector for food security and economy of any country. Crop production is one of the most important components of agriculture. The development of high yielding varieties of crops and the improved management practices in the last 4 decades resulted in higher production through improving the productivity of crops.

A large number of books are available on crop production. Even though they cover all the aspects of crop production of almost all the crops, only a few books cover as per the syllabus of most of the Universities. The students are finding difficult to refer these books as these books are not according to the prescribed syllabus. A need was felt and expressed by the students to bring out a book in a simple manner incorporating all concepts and later advances in crop production.

The present book Crop Production Technology - (*kharif* crops) is an earnest attempt in this direction. This publication is prepared in a simple language so that all the students can be benefited. This publication is also prepared as per the syllabus of ICAR and followed by most of the state Agricultural Universities as full and with little modification in some other Universities. Hope this book will cater the needs of the students and teachers of various disciplines.

We appreciate the encouragement and support extended by the scientists, family members who endured many things during the course of preparation of this edition.

Authors

Contents

Foreword ... v

Preface ... vii

1. Importance, Area, Production and Productivity of Major Cereals and Millets of India and Tamil Nadu ... 1

2. Importance, Area, Production and Productivity of Pulses and Oilseeds Crops of India and Tamil Nadu ... 9

3. Rice-Origin-Geographic Distribution- Economic Importance-Varieties-Soil and Climatic Requirement ... 15

4. Rice - Cultural Practices - Yield Economic Benefits ... 21

5. Special Type of Rice Cultivation SRI and Hybrid Rice Cultivation ... 35

6. Maize - Origin, Geographic Distribution, Economic Importance Soil and Climatic Requirement Varieties, Cultural Practices and Yield ... 43

7. Sorghum and Pearl Millet-Origin, Geographic Distribution, Economic Importance, Soil and Climatic Requirement, Varieties, Cultural Practices and Yield ... 51

8. Finger Millet and Minor Millets - Origin, Geographic Distribution, Economic Importance, Soil and Climatic Requirement, Varieties, Cultural Practices and Yield ... 65

9. Pigeonpea - Origin, Geographic Distribution, Economic Importance, Soil and Climate Requirement and Yield ... 79

10. Greengram, Blackgram & Cowpea-Origin,Geographic Distribution, Economic Importance, Soil and Climatic Requirement, Varieties, Cultural Practices and Yield - Agronomy of Rice Fallow Pulses ... 89

11. Groundnut-Origin, Geographic Distribution, Economic Importance, Soil and Climatic Requirements-Varieties, Cultural Practices and Yield 105

12. Sesame, Soybean - Origin Geographical Distribution Economic Importance, Soil and Climatic Requirements, Varieties Cultural Practices and Yield 119

13. Cotton - Origin, Geographic Distribution, Economic Importance Soil and Climatic Requirement, Varieties, Cultural Practices and Yield 133

14. Jute - Origin, Geographical Distribution, Economic Importance Soil and Climatic Requirements Varieties, Cultural Practices and Yield 151

15. Fodder Sorghum - Origin, Geographic Distribution, Economic Importance Soil and Climatic Requirement Varieties, Cultural Practices and Yield and Fodder Preservation 159

16. Napier-Bajra Hybrids - Origin, Geographic Distribution, Economic Importance, Soil and Climatic Requirement, Varieties, Cultural Practices and Yield 167

17. Fodder Cowpea, Cluster Bean - Origin, Geographic Distribution Economic Importance, Soil and Climatic Requirement, Varieties Cultural Practices and Yield and Fodder Preservation 173

1

Importance, Area, Production and Productivity of Major Cereals and Millets of India and Tamil Nadu

Cereals

The word cereal was derived from **Ceres**, the name of the Roman goddess of harvest and agriculture. Cereals are grasses (members of the monocot family Poaceae, also known as Gramineae) cultivated for the edible components of their grain (botanically, a type of fruit called a caryopsis), composed of the endosperm, germ and bran. Cereal grains are grown in greater quantities and provide more food energy worldwide than any other type of crop and therefore they are called staple crops.

In their natural form (as in whole grain), they are a rich source of carbohydrates, vitamins, minerals, fats, oils and protein. However, when refined by the removal of the bran and germ, the remaining endosperm is mostly carbohydrate and lacks the majority of other nutrients. In some developing nations, grain in the form of rice, wheat, millet or maize constitutes a majority of daily sustenance. In developed nations, cereal consumption is moderate and varied but still substantial. Globally, more than 2000 m. t of cereals are produced from about 700 m. ha with an average productivity of about 3000 kg/ha.

Millets "Miracle grains"

Millets share a set of characteristics which make them unique amongst cereals. Millets are drought resistant and require few external inputs. They can be grown under harsh circumstances in arid and semi-arid environments requiring less water than many other cereals and are often able to cope with poor soils. For this they are sometimes called 'miracle grains' or 'crops of the future'. Millets provide food and livelihood security to millions of households, in particular, to small and marginal farmers and inhabitants of rainfed areas, especially in remote tribal areas. Millets are usually cultivated as dual-purpose crops providing both food grain for human consumption and straw for animals, contributing to economic efficiency in mixed farming systems. In addition, millets sequestrate carbon, thereby adding to CO_2 abatement opportunities, contributing to improved agro-biodiversity by their rich varietal diversity, allowing for mutually beneficial intercropping with other vital crops and have significant cultural value due to their long history on the Indian subcontinent.

Millets have been cultivated for around 3,000 years making them an integral part of the culture and history of India. References to millets can be found in mythology, poetry, religious practices, ayurvedic recipes and in numerous dishes. Millets are not only food grains; they are still intricately interwoven in the socio-cultural fabric of numerous regions.

They are small grained cereals. They are food for millions of people and so referred as poor man's cereals because of choice next to rice and wheat. Most of the cereals are grown in Asia, Africa and Russia.

Specialty of millets

Under infertile soil, intense heat and scanty rain, they produce greater quantity of grain than any other crop. They require only a short growing period. When the season fails for other crops, they are preferred to catch the remaining season yet they can yield satisfactory net return. They are valuable as second and catch crop. The seed requirement is low. They are important for counties where People: Land Ratio is wide.

They are warm weather annual grasses. Height ranges from 30-130 cm. They are all short-day plants. Millets are grouped as major and minor millets based on area, production productivity and importance.

Uses of millets

They are staple food for drier regions of the world. Used for making bread and cakes. Mineral contents of millets are generally higher. They are good source of fiber. Whole grain furnishes more thiamine than milled rice.

Area, production, productivity of cereals and millets in India (Ranking)

Crop	Rank	Area	Production	Productivity
Rice	I	West Bengal	West Bengal	Punjab
	II	Uttar Pradesh	Punjab	Tamil Nadu
	III	Odisha	Andhra Pradesh	Andhra Pradesh
Maize	I	Karnataka	Andhra Pradesh	Tamil Nadu
	II	Rajasthan	Karnataka	West Bengal
	III	Madhya Pradesh	Maharashtra	Andhra Pradesh
Sorghum	I	Maharashtra	Maharashtra	Madhya Pradesh

	II	Karnataka	Karnataka	Andhra Pradesh
	III	Rajasthan	Madhya Pradesh	Gujarat
Pearl millet	I	Rajasthan	Rajasthan	Uttar Pradesh
	II	Maharashtra	Uttar Pradesh	Haryana
	III	Uttar Pradesh	Haryana	Tamil Nadu

Area, production and yield of cereals in India

Area - Million Hectares, Production - Million Tonnes, Yield - Kg /ha

Year	Rice			Maize		
	A	P	Y	A	P	Y
2001-02	44.90	93.34	2079	6.58	13.16	2000
2002-03	41.18	71.82	1744	6.64	11.15	1681
2003-04	42.59	88.53	2078	7.34	14.98	2041
2004-05	41.91	83.13	1984	7.43	14.17	1907
2005-06	43.66	91.79	2102	7.59	14.71	1938
2006-07	43.81	93.36	2131	7.89	15.10	1912
2007-08	43.91	96.69	2202	8.12	18.96	2335
2008-09	45.54	99.18	2178	8.17	19.73	2414
2009-10	41.92	89.09	2125	8.26	16.72	2024
2010-11	42.86	95.98	2239	8.55	21.73	2542
2011-12	44.01	105.30	2393	8.78	21.76	2478
2012-13	42.75	105.24	2462	8.67	22.26	2566
2013-14	43.95	106.54	2424	9.43	24.35	2583
2014-15	44.13	105.48	2391	9.18	24.17	2632
2015-16	43.38	104.32	2404	8.80	22.56	2563
2016-17	43.99	109.69	2494	9.63	25.89	2689
2017-18	44.25	112.75	2576	9.38	28.75	3065
2018-19	44.16	116.48	2638	9.03	27.70	3070
2019-20	43.78	118.87	2705	9.57	28.77	3006

Area, production and yield of millets in India

Year	Sorghum			Pearl millet			Finger millet		
	A	P	Y	A	P	Y	A	P	Y
2001-02	9.80	7.56	771	9.53	8.28	869	1.646	2.37	1442
2002-03	9.30	7.01	754	7.74	4.72	610	1.415	1.32	930
2003-04	9.33	6.68	716	10.61	12.11	1141	1.666	1.97	1180
2004-05	9.09	7.24	797	9.23	7.93	859	1.552	2.43	1567
2005-06	8.67	7.63	880	9.58	7.68	802	1.533	2.35	1534
2006-07	8.47	7.15	844	9.51	8.42	886	1.177	1.44	1226
2007-08	7.76	7.93	1021	9.57	9.97	1042	1.387	2.15	1552

2008-09	7.53	7.25	962	8.75	8.89	1015	1.381	2.04	1552
2009-10	7.79	6.70	860	8.90	6.51	731	1.268	1.89	1489
2010-11	7.38	7.00	949	9.61	10.37	1079	0.879	2.19	1705
2011-12	6.25	5.98	957	8.78	10.28	1171	0.768	1.93	1641
2012-13	6.21	5.28	850	7.30	8.74	1198	0.739	1.57	1396
2013-14	5.82	5.39	926	7.89	9.18	1164	0.765	1.98	1661
2014-15	6.16	5.44	884	7.31	9.18	1255	1.208	2.06	1706
2015-16	6.07	4.23	697	7.12	8.06	1132	1.138	1.82	1601
2016-17	5.62	4.56	812	7.45	9.72	1305	1.016	1.38	1363
2017-18	5.24	4.80	956	7.48	9.21	1231	1.19	1.98	1662
2018-19	5.02	3.48	849	7.48	8.66	1219	0.89	1.24	1390
2019-20	4.09	4.77	989	7.11	10.36	1374	1.00	1.76	1747

Source: www.indiastat.com/agriculture/production of crops

Area, production and yield of small millets in India

Area - Million Hectares, Production - Million Tonnes, Yield - Kg /ha

Year	Small millets		
	A	P	Y
2001-02	3.087	1.111	360
2002-03	3.015	0.867	288
2003-04	3.548	1.702	480
2004-05	3.341	1.058	317
2005-06	3.109	0.946	304
2006-07	3.194	1.115	349
2007-08	3.222	1.523	409
2008-09	2.483	1.035	364
2009-10	3.091	0.707	229
2010-11	3.610	1.862	516
2011-12	3.387	0.707	483
2012-13	2.719	1.862	436
2013-14	3.383	1.634	475
2014-15	3.019	1.503	498
2015-16	3.827	1.592	416
2016-17	4.326	2.165	500
2017-18	5.462	4.389	804
2018-19	0.453	3.330	734
2019-20	0.458	3.700	809

Area, production and yield of cereals and millets in Tamil Nadu

Area – Hectares, Production – Tonnes, Yield - Kg / ha

Year	Rice			Maize			Sorghum		
	A	P	Y	A	P	Y	A	P	Y
2000-01	2059878	6583630	3196	81000	140000	1717	317233	274840	866
2001-02	1516537	3577108	2359	73000	118000	1624	319607	210793	660
2002-03	1396651	3222776	2308	121057	191646	1583	401607	245933	612
2003-04	1872822	5061622	2703	118439	176381	1489	376739	252063	669
2004-05	2050455	5209433	2541	189893	294717	1552	316274	231449	732
2005-06	1931000	6611000	3423	202830	241217	1189	294000	294000	999
2006-07	1789000	5040000	2817	215767	231462	1072	284000	248000	874
2007-08	1931603	5183385	2682	223428	810057	3626	258876	213436	824
2008-09	1845553	5665258	3070	286639	1257889	4388	238476	221960	931
2009-10	1905726	5792415	3039	244159	1138126	4661	243465	246981	1014
2010-11	1903772	7458657	3918	230489	1027536	4458	197696	252522	1277
2011-12	1493276	4050334	2712	280629	1695467	6042	210893	174966	830
2012-13	1725730	7115195	4123	291052	946363	3252	347131	513313	1479
2013-14	2059878	6583630	3196	380429	2245216	5902	317233	274840	866
2014-15	1795000	5727800	3191	322000	2067900	6423	347500	512700	1475
2015-16	2000200	7517100	3758	355100	2489300	7010	339200	468600	1380
2016-17	1442800	2369400	1642	315030	9533800	3026	268390	153870	573
2017-18	1828910	6638860	3630	324520	259168	7986	385640	430590	1117
2018-19	1721000	6130000	3562	390000	283400	7258	385000	464000	1204
2019-20	1907000	7171000	3760	333000	247600	7424	450000	520000	1156

Source: www.agricoop.nic.in

Area, Production and Yield of millets in Tamil Nadu

Area – Hectares, Production – Tonnes, Yield - Kg /ha

Year	Pearl millet			Finger millet			Foxtail millet		
	A	P	Y	A	P	Y	A	P	Y
2001-02	102020	88682	869	104286	140169	1344	1000	500	507
2002-03	158851	172341	1085	118439	176381	1489	2210	1091	494
2003-04	97608	124300	1273	108845	154085	1416	1626	793	488
2004-05	81925	94799	1157	99549	132172	1325	1355	665	491
2005-06	66000	99000	1511	95000	148000	1511	832	513	617
2006-07	60000	86000	1436	94000	176000	1436	864	405	864
2007-08	56672	84021	1483	90079	169944	1887	813	411	813

2008-09	54427	82652	1519	82335	160939	1955	1074	517	467
2009-10	49482	77369	1564	75650	171096	2262	863	391	471
2010-11	46664	114447	2453	82805	224862	2716	689	330	478
2011-12	42928	56505	1316	70294	138011	1963	798	375	453
2012-13	54412	117436	2158	118699	362343	3053	1020	480	481
2013-14	125093	152970	1223	124958	235310	1883	1411	652	251
2014-15	57700	177570	3077	104400	349600	3348	-	-	-
2015-16	51600	141750	2747	90000	271200	3013	-	-	-
2016-17	49670	102260	2059	61360	114430	1865	-	-	-
2017-18	63300	143520	2277	86510	321300	3714	-	-	-
2018-19	46800	118000	2517	78600	256000	3257	-	-	-
2019-20	67400	185000	2743	84500	274000	3247	-	-	-

Source: www.agricoop.nic.in

Area, production and yield of millets in Tamil Nadu

Area – Hectares, Production – Tonnes, Yield - Kg /ha

Year	Kodo millet			Little millet		
	A	P	Y	A	P	Y
2001-02	14200	15800	1114	42900	31000	723
2002-03	7748	9876	1275	48126	27808	578
2003-04	8547	10455	1223	33819	27866	824
2004-05	5871	6676	1137	27577	19604	711
2005-06	11056	30666	2774	24997	25750	1030
2006-07	5546	8144	5546	24229	24281	1002
2007-08	4086	5679	4086	21231	16503	777
2008-09	5930	8805	1485	22292	19682	883
2009-10	8039	12946	1610	17953	17250	961
2010-11	4156	8362	2012	20378	25060	1230
2011-12	3340	4511	1351	17423	19071	1095
2012-13	3570	6187	1734	22348	25445	1139
2013-14	11345	16330	1428	48076	33060	688

Source: www.agricoop.nic.in

Questions

I. Choose the best from the choices given:

1. The state having the highest area and production of sorghum is

   a) Maharastra b) Uttar Pradesh

   c) Madhya Pradesh d) Karnataka

2. The state having the highest productivity of maize is

   a) Karnataka b) Andhra Pradesh

   c) Tamil Nadu d) Punjab

3. The state having the highest area and production of pearl millet is

   a) Karnataka b) Andhra Pradesh

   c) Rajasthan d) Uttar Pradesh

4. Millets are

   a) Long day plants b) Short day plants

   c) Day neutral plants d) None

5. Millets have been cutivated for around

   a) 1000 years b) 3000 years

   c) 2000 years d) 4000 years

II. Fill in the blanks

1. The word cereal was derived from __________, the roman goddess of harvest
2. Cereals are botanically, a type of fruit called a __________
3. ____________ provide food and livelihood security to inhabitants of rainfed areas, especially in remote tribal areas.
4. Millets are called as ___________
5. Globally cereals are grown in _________ m ha

III. Write short answers

1. Importance of cereals
2. Nutrients in cereals
3. Importance of millets
4. Speciality of millets
5. Uses of millets

2

Importance, Area, Production and Productivity of Pulses and Oilseeds Crops of India and Tamil Nadu

Pulses

Pulses may be defined as the dried edible seeds of cultivated legumes. They belong to the family of peas, beans and lentils (Family: Fabaceae). English word pulse is taken from the Latin word 'Puls', meaning pouage or thick pap. Pulses are a large family and various species are capable of surviving in very different climates and soils. Pulses are cultivated in all parts of the world and they occupy an important place in human diet.

Pulses constitute an important group of food in the world. They are the most important source of proteins. Pulses contribute substantially to food production system by enriching the soil through biological nitrogen fixation and improving soil physical conditions. They are rightly called 'unique jewels' of Indian crop husbandry.

Pulses contain more protein than any other plant. They serve as a low-cost protein to meet the needs of the large section of the people. They have, therefore, been justifiably described as 'the poor man's meat'. Their low moisture content and hard test or seed-coat permits storage over long periods. In addition to providing dry pulses, many of the crops are grown for their green edible pods and un-ripe seeds.

In general, pulses contain 20 to 28% protein. Their carbohydrate content is about 60%. Pulses are also fairly good sources of thiamin and niacin and provide calcium, phosphorus and iron.

Oilseeds

Oilseeds are important as are the pulses in the country. The principal oilseeds include groundnut, rapeseed and mustard seeds. While the former is a *Kharif* crop, depending wholly upon reasonable but timely rainfall, the latter is a *Rabi* crop, fundamentally confined only to nonirrigated areas. As a result, their production as well as productivity is subject to climatic variations and market hypotheses. The other oilseeds includes sesame, linseed, castor, safflower, soybeans, sunflower, cotton seeds and copra. Rapeseed and mustard seeds belong to the wheat belt of north and central India. Groundnut, on the other hand, is grown in west and south India. Gujarat is the dominant producer of groundnut. While population has been mounting at 2% per annum, the demand for oil has been rising at 5% every year.

Area, production and yield of pulses in India

Area - Million Hectares, Production - Million Tonnes, Yield - Kg /ha

Year	Pigeon pea			Black Gram			Greengram		
	A	P	Y	A	P	Y	A	P	Y
2001-02	3.33	2.26	679	3.30	1.50	454	3.09	1.11	360
2002-03	3.36	2.19	651	3.33	1.47	423	3.01	0.87	288
2003-04	3.52	2.36	670	-	1.47	-	3.55	1.71	475
2004-05	3.52	2.35	667	3.33	1.33	-	3.34	1.06	317
2005-06	3.58	2.74	765	2.99	1.25	419	3.11	0.95	304
2006-07	3.56	2.31	650	3.10	1.44	473	3.19	1.12	349
2007-08	3.73	3.08	826	3.23	1.46	462	3.73	1.52	409
2008-09	3.38	2.27	671	2.71	1.17	444	2.84	1.03	364
2009-10	3.47	2.46	711	2.99	1.24	422	3.07	0.69	225
2010-11	4.37	2.86	654	3.26	1.76	543	3.55	1.80	512
2011-12	4.01	2.65	661	3.30	1.77	555	3.43	1.63	498
2012-13	3.89	3.02	776	2.29	1.90	500	2.75	1.19	436
2013-14	3.88	3.29	849	-	1.70	-	3.38	1.61	474
2014-15	3.85	2.80	729	3.24	1.94	604	3.01	1.50	498
2015-16	3.96	2.56	913	3.62	1.95	537	3.82	1.59	416
2016-17	5.33	4.87	646	4.47	2.83	632	4.32	2.16	500
2017-18	4.44	4.23	967	5.23	3.49	662	4.24	2.02	477
2018-19	4.55	3.32	729	5.60	3.06	546	4.75	2.46	516
2019-20	4.53	3.89	859	4.53	2.08	459	4.58	2.51	547

Source: www.indiastat.com/agriculture/production of crops

Area, production and yield of oilseeds in India

Area - Million Hectares, Production - Million Tonnes, Yield - Kg /ha

Year	Ground nut			Sesame			Soybean		
	A	P	Y	A	P	Y	A	P	Y
2001-02	6.24	7.03	1127	1.4444	0.698	418	6.34	5.96	940
2002-03	5.94	4.12	694	1.7003	0.441	306	6.11	4.65	762
2003-04	5.99	8.13	1357	1.844	0.782	460	6.55	7.82	1193
2004-05	6.64	6.77	1020	1.7232	0.674	366	7.57	6.87	908
2005-06	6.74	7.99	1187	1.7032	0.641	372	7.71	8.27	1073
2006-07	5.62	4.86	866	1.7991	0.618	363	8.33	8.85	1063
2007-08	6.29	9.18	1459	1.8091	0.757	421	8.88	10.97	1235
2008-09	6.16	7.17	1163	1.9421	0.640	354	9.51	9.91	1041
2009-10	5.48	5.43	991	2.0832	0.588	303	9.73	9.96	1024
2010-11	5.86	8.26	1411	1.9015	0.893	429	9.60	12.74	1327
2011-12	5.26	6.96	1323	1.7058	0.810	426	10.11	12.21	1208
2012-13	4.72	4.70	995	1.6789	0.685	402	10.84	14.67	1353
2013-14	5.53	9.67	1750	1.7756	0.715	426	12.20	11.99	983
2014-15	7.76	7.40	1552	1.7461	0.827	474	10.91	10.37	951
2015-16	4.59	6.73	1495	1.9509	0.850	426	11.60	08.57	738
2016-17	5.33	7.46	1398	1.6669	0.747	448	11.18	13.15	1177
2017-18	4.44	4.23	967	5.23	3.49	662	4.24	2.02	477
2018-19	4.73	6.73	1422	1.42	0.689	485	11.13	13.27	1192
2019-20	4.83	9.95	2063	1.62	0.658	405	12.19	11.23	921

Source: www.indiastat.com/agriculture/production of crops

Area, production and yield of pulses in Tamil Nadu

Area – Hectares, Production – Tonnes, Yield - Kg /ha

Year	Red gram			Green gram			Black gram		
	A	P	Y	A	P	Y	A	P	Y
2001-02	44127	24067	545	112812	48053	426	196888	78555	399
2002-03	44914	27475	612	125690	53315	424	185736	75920	409
2003-04	43416	28979	667	154959	61760	399	226364	82998	367
2004-05	37769	20400	540	136699	45881	336	215448	70758	328
2005-06	29127	21334	732	134071	77404	577	251014	143053	570
2006-07	30057	21077	701	158691	46213	291	307515	79980	260
2007-08	27483	16703	608	138598	31336	226	263671	82983	315

2008-09	26558	20274	765	138138	47673	345	259722	98712	380
2009-10	35751	23671	662	171666	57683	336	304432	123011	404
2010-11	35968	31292	870	164069	85118	519	308263	178816	580
2011-12	42065	33105	787	118615	33674	284	208625	88706	425
2012-13	59643	57666	967	195285	151400	775	365128	310658	851
2013-14	63613	41420	651	128995	53470	415	266123	104320	392
2014-15	72400	77000	1064	230000	181000	787	373800	358900	760
2015-16	56560	37050	655	238800	125100	524	395200	264120	668
2016-17	59010	53580	908	165970	561000	338	429780	273960	637
2017-18	49230	53810	1093	180590	782400	433	426340	301570	707
2018-19	41300	51600	1248	170300	76800	451	440900	274400	622
2019-20	41800	53200	1273	171900	76400	445	405300	317300	783

Source: www.agricoop.nic.in

Area, production and yield of oil seeds in Tamil Nadu

Area – Hectares, Production – Tonnes, Yield - Kg /ha

Year	Ground nut			Sesame		
	A	P	Y	A	P	Y
2001-02	502090	717403	1429	83848	45990	548
2002-03	591696	918241	1552	64481	28047	435
2003-04	615877	1005342	1632	83835	29004	346
2004-05	618835	1097592	1775	72725	33840	465
2005-06	508032	1006475	1981	65118	30772	469
2006-07	535211	1047586	1957	52624	27328	519
2007-08	489844	974768	1990	74376	32201	433
2008-09	413011	895718	2169	63693	32242	506
2009-10	385509	895638	2323	62677	29021	463
2010-11	385612	1060654	2751	48189	25387	527
2011-12	339361	785362	2314	43175	26447	613
2012-13	336621	915884	2721	33181	17179	518
2013-14	662982	1249630	1885	56591	33708	596
2014-15	336500	926400	2753	64200	44800	697
2015-16	346600	892300	2574	46300	29400	634
2016-17	282490	588850	2084	28230	10840	384
2017-18	327350	1007530	3078	41580	23080	555
2018-19	335000	911000	2718	44610	24450	548
2019-20	346000	1033000	2980	53010	36470	688

Source: www.agricoop.nic.in

Questions

I. Choose the best from the choices given:

1. Seeds of leguminous plants used as food are called as
   a) Pulses   b) Oilseeds
   c) Dhal   d) Vegetables
2. The state having the highest area and production in pigeon pea is
   a) Maharastra   b) Uttar Pradesh
   c) Madhya Pradesh   d) Punjab
3. The state having the highest productivity of blackgram is
   a) Maharashtra   b) Andhra Pradesh
   c) Tamil Nadu   d) Punjab
4. The state having the highest area and production of greengram is
   a) Karnataka   b) Madhya Pradesh
   c) Rajasthan   d) Uttar Pradesh
5. The dominant producer of groundnut is
   a) Tamil Nadu   b) Andhra Pradesh
   c) Gujarat   d) Maharashtra

II. Fill in the blanks

1. ________are the most important source of proteins for human population
2. The poor man's meat is __________
3. Pulses contain ____________ % protein.
4. The state having the highest productivity of pigeon pea is ______.
5. Oilseed demand is increasing every year by _________%

III. Write short notes

1. Importance of pulses
2. Important pulse crops grown in India
3. Importance of oilseeds
4. Important oilseed crops grown in India
5. Important pulses and oilseeds grown in Tamil nadu

3

Rice-Origin-Geographic Distribution- Economic Importance-Varieties-Soil and Climatic Requirement

Cereals

Cereals are the cultivated grasses grown for their edible starchy grains. Cereals are larger grains used as staple food – Rice, wheat, maize, barley, oats *etc.*

Cereal grain contains 60 to 70% of starch and is excellent energy rich foods for humans. In almost every country and region, cereals provide the staple food. In the world as a whole, only 5% of starchy staple food comes from root crops (mainly cassava, potato, and yams, depending on climate), whereas the rest is from cereal. Cereals are an excellent source of fat-soluble vitamin E, which is an essential antioxidant. Whole cereal grains contain 20 to 30% of the daily requirements of the minerals such as selenium, calcium, zinc and copper.

Rice (*Oryza sativa*)

Rice farming is the largest single use of land for food. Rice production totaled 600 million tonnes. Ninety per cent of rice is produced in Asia alone. Only 6-7% of production is exported from area of production. Rice field covers 11% of arable land. It is the most important economic activity on earth. Rice eaters and growers form the bulk of the worlds' poor. It is the single most important activity of rural people in the world. Rice is grown in 250 million Asian farms. Rice farming is 10,000 years old. It is the staple food for the largest number of humanities in the world. It is the single largest source of energy for poor. Rice is synonym with food throughout Asia.

Rice is the staple food of over 3 billion people in Asia, which accounts for the production and consumption of 90% of world rice.

India is the major rice producer next to China. In India, rice accounts for 22% of the total cropped area, 39% of the total area under cereals and 31% of the total area under food grains. Rice production is roughly 46% of the total production of cereals and 41% of the total grain output. Rice is cultivated in an area of 43.8

million ha with a production of 112.8 million tonnes. The irrigated area of rice is only 45%. The average yield of rice in India is only 2576 kg/ha whereas it is 6000 kg in Korea, 5800 kg in Japan, 5700 kg in China and 5600 kg in Egypt. In our country Uttar Pradesh, West Bengal, Bihar, Orissa, Madhya Pradesh and Assam cover about two third of rice area with only 50% of the total rice production.

Origin: South East Asia

1. Vavilo (1930) considered South West part of the Himalayas as the primary centre of origin for Rice.
2. Ramiah (1953) reported that India and Indo–China were the countries where cultivated rice arose because of the presence of wild species, the varietal diversity of cultivated rice and the presence of many dominant genes.

Distribution

The leading countries producing rice crop are Japan, Brazil, China, India, Indonesia, Bangladesh, Vietnam, Thailand, Myanmar and Philippines.

In India, rice is grown in almost all the states. Andhra Pradesh, Bihar, Uttar Pradesh, Madhya Pradesh and West Bengal lead in the area. West Bengal and Uttar Pradesh have the highest rice production. The average yield per hectare is highest in Punjab (3346 kg/ha).

Economic importance of rice and by-products

Apart from rice as food (parboiled rice and raw rice) it is used in the following ways.

1. **Parched rice (Arisipori)**
2. **Flaked rice (Aval)**
3. **Sake:** An alcoholic beverage prepared by fermentation of rice in water. This is an important alcoholic beverage of Japanese.
4. **Snacks:** Papads, vermicelli, crispies, murukku, ribbon pakoda
5. **Bran:** By product of rice milling used as cattle feed.
6. **Bran oil:** Bran oil is extracted immediately after milling the rice by hydraulic pressure or with solvents. Bran oil contains 3-9% wax, which is removed by filtering or centrifuging. It is used as an edible oil.

7. **Bran wax:** By product of bran oil extraction contains esters of waxy acid with higher alcohols used in chocolate industry, coating for candy, cosmetics etc.,
8. **Husk:** Paddy husk is used as fuel mainly in rural areas in brick kilns and compost making.
9. **Straw:** Used for feeding cattle, preparing straw boards, straw pulp in paper industry, thatching, hats, mats, rope and baskets and mushroom cultivation.
10. Ground rice meal from broken grain is used in confectionery.

Industrial uses

11. **Starch:** Over boiled rice is used for starching clothes. The starch is used in foods, cosmetics and textile manufacturing.
12. **Flour:** Rice flour is used as a thickening agent for white sauces, gravies and puddings.
13. **Beverages:** Various alcoholic beverages are manufactured from rice.
14. **Medicines:** Rice is used in various forms as a medicine.

Climate

The temperature range of 30.4 to 32.6°C from transplanting to harvest is ideal. The growing season should be free from frost, extreme cold, severe drought, heavy winds, cyclones and with bright sunshine during ripening phase and harvest period.

Soil

The soil type should be loamy or alluvial with a pH range of 5.0 to 8.0 without salinity and alkalinity. The most suited soils are heavy soils, clays or clay loams. An abundant supply of water for irrigation and good drainage facilities are needed.

Varieties of Tamil Nadu

a) Short duration (100-120 days)

ADT 36, ADT 37, IET 1444, TKM 9, ASD 16, ASD 17, ASD 18, ASD 20, IR 64, PMK 1, PMK 2, MDU 1, MDU 5, JJ 92, TPS 1, TKM 11, CO 47, CO (R) 51, ADT 42, ADT 43, CORH 1, CORH 3, CORH 4, ADTRH 1, ADTRH 2, ADTRH 3, SJR (IET-19972), MDU 6, TPS 5, ADT (R) 45, Anna (R) 4, CORH 3, TRY3, PKM (R) 3,

b) Medium duration (120 – 140 days)

IR 20, IR 36, Bhavani, ADT 38, ADT 39, ADT 40, ADT 49, MDU 3, MDU 4, Paiyur 1, CO 43, CO 45, TPS 2, TPS 3, CO 46, CO 47, CO 50, CO 52, TKM 13, TRY 3, White Ponni, ASD 19, TKM 10, TRY 1, CORH 2, CORH 4, ADT (R) 46, Paiyur 1, Arize Tej (HRI-169), (IET 21411) (Hybrid), DRR Dhan 45 (Zinc rich variety), CR Dhan 310 (Protein rich variety), CR Dhan 200 (aerobic), BPT 5204, 28P 0 9, JKRH 3333, TKM 13, DRRH 2 & 3.

c) Long duration: (i) 140 – 160 days (ii) Above 160 days

Paiyur 1, Ponmani (CR 1009) (single crop samba season), ADT 51, ADT 50, ADT 40 (shallow deep-water areas), AU 2, Savithiri, PY 4 (Jawahar), CR Dhan 801, CR 1009 sub-1.

Latest varieties: CO(R) 48, CO(R) 49, CO(R) 50, TRY 2, ADT 53, VGD 1, Pusa Basmati 1637, Pusa Basmati 1728, IR 64 Drt I, Arize 6444 Gold, US 305, DRR Dhan 40, 27P31, NK 5251, VNR 2355 PLUS

Hybrids: CORH 1, ADTRH 1, CORH 2, ADTRH 2, CORH 3.

Short duration rice has no lag phase since maximum tillering and flowering occurs simultaneously unlike medium and long duration rice varieties.

Seasons:

National Level

Major season	Sowing	Harvesting
Kharif	June-July	Nov-December
Autumn	May-June	Sep-October
Rabi	Nov-December	March-April

Rice seasons and varieties of Tamil nadu

S.No.	Season	Varieties	Period	Remarks
1.	Sornavari	TKM9, ASD16, 17, 18, ADT37, IR64, ADT41	April-May	First season, short duration varieties
2.	Kar	ADT36, ASD16, SD17, ASD18, IR50, IR64, MDU 5, ADT 43, CO 47, ADT (R) 45, ADT (R) 47, CORH 3	May-June	First season, short duration varieties
3.	Kuruvai	IR50, TKM9, ADT36, IR64, ADT37, ASD16, ASD18MDU5, IR 50, ADT 43, ADT (R) 45, ADT (R) 47, ADT (R) 48, CO 47, CORH 3	June-July	First season, short duration varieties

4.	Samba	Ponmani, IR20, CO43, ADT39, ADT40, MDU4, Paiyur1, White PonniCR 1009 sub-1, ADT 38, ADT (R) 44, CORH 4, CO (R) 48, CO (R) 49, CO (R) 50, ADT (R) 49, TNAU Rice ADT 50, Swarna sub1	July-August	Medium and long duration varieties if water available
5.	Late Samba	ADT39, CO43, White Ponni ADT (R)46, CO (R) 48, CO (R) 49, CO (R) 50, ADT (R) 49	Sep-Oct	Medium duration varieties
6.	Thaladi	MDU 2, 3, 4, ADT38, 39, IR64, CO43, CO45, ASD16, 18, Ponmani, White Ponni ,CR 1009 sub 1, CO 43, TPS 2, TPS 3, ADT (R) 44, ASD 19, MDU 5, ADT (R) 46, CORH 4, CO (R) 48, CO (R) 49, CO(R) 50, ADT (R) 49	Sep-Oct	Medium duration varieties
7.	Pishanam	White Ponni, TPS2, ASD18, IR20, CO45, ADT39, 40	Sep-Oct	Medium duration varieties
8.	Late Pishanam	ADT37, ASD16, 17, 18	Oct-Nov	Short duration varieties
9.	Navarai or kodai	IR50, ADT37, ADT39, ASD16, 18, IR64, ADT 36, ASD 18, ASD 19, ADT 42, MDU 5, ASD 20, CORH 3, ADT 43, ADT 45	Dec-Jan	Short duration varieties

Questions

I. Choose the best from the choices given

1. The continent which accounts for the production and consumption of 90% of world rice is
   a) America b) Australia
   c) Asia d) Africa
2. In Asia, the major rice producer next to China is
   a) Japan b) India
   c) Thailand d) Philippines
3. The state having the highest productivity of rice is
   a) West Bengal b) Andhra Pradesh
   c) Punjab d) Uttar Pradesh
4. The most suited soil for rice is
   a) Sandy soil b) Light soil
   c) Red soil d) Clay soil

5. White Ponni is an important variety for the season

a)	Sornavari	b)	Kuruvai
c)	Samba	d)	Navarai

II. Fill in the blanks

1. Vavilov considered Southwest part of __________ as the primary center of origin of rice.
2. Rice is the staple food of over 3 billion people in __________.
3. Cereal grain is rich in _________
4. Rice production in India is ______ m t.
5. The average yield of rice in India is _____ kg ha^{-1}.

III. Write short notes

1. Rice seasons of India
2. Short duration rice varieties of Tamilnadu
3. Economic importance of rice
4. Climatic requirement of rice
5. Industrial uses of rice.

4

Rice - Cultural Practices - Yield Economic Benefits

Rice Nursery: Methods of nursery

1. Wet bed method

The field is ploughed once and harrowed twice or thrice until the soil becomes well puddle. Beds are made slightly raised of 1 – 1.5 m in width and conveniently large in size and drainage channels in between the beds. The total area of nursery required to plant one hectare of land is 20 cents (800 m^2). This method is suited where assured irrigation facilities are available. The beds should not be allowed to dry because they develop cracks and the seedlings will get spoiled. This method keeps down the weeds in the nursery. Seedlings are transplanted at 21-25 and 25-30 days age for SDV and MDV, respectively. Sparse sowing is recommended in case of hybrid rice. One kg of hybrid seed is sown in one cent area of nursery. This helps in good tillering and robust seedlings and makes it possible to transplant single seedling per hill. Seedlings are planted at 20-25 days.

2. Dry bed method

This method is adopted in high rainfall areas having no irrigation facilities. Light soils are chosen. The land is ploughed, harrowed and leveled (but never puddled). Raised beds of 8-10 cm height, 1-1.5 m width and 8-10 m length are made with 30 cm wide drainage channels between the beds. Fifty to sixty beds are required to plant one-hectare main field area.

3. Dapog method

Seedlings are raised without contact with soil which are ready for transplanting on 12th day. Well compacted seed beds, concrete floor, wooden planks or trays can be used and the surface is covered with polythene sheets. The sprouted seeds are broadcasted uniformly @ 1.5 kg per sq. m area. The beds are moistened constantly and pressed slightly two to three times a day with a smooth wooden plank for the first three days, so that the roots remain in contact with water.

When the seedlings attain 2 cm height a constant film of water should be maintained. On 12th day, the nursery can be cut into strips and rolled like a mat and then transported to the planting site easily. Seedlings of one sq.m area can be transplanted in 200 sq.m area.

Nursery Management

SEED RATE

Duration	:	Seed rate
Short duration	:	60 kg/ha
Medium duration	:	40 kg/ha
Long duration	:	30 kg/ha
Hybrids	:	20 kg/ha

Pre-treatment of seeds with nutrients

The seed should be soaked in water for 10 hrs and the excess water drained. If the seeds are required for immediate sowing, the soaked seed should be kept in gunny in dark and covered with extra gunnies for 24 hours for sprouting. To induce tolerance under short and prolonged drought situation in *kharif* season in Periyar-Vaigai command, a combination of seed treatment with 1% KCl + CCC at 500 ppm with foliar application at vegetative stage is effective in mitigating the drought and increasing the yield.

Pre-treatment of seeds with fungicides

a) *Dry seed treatment*

Any one of the fungicides *viz*., thiram, captan, carbendazim should be mixed at the rate of 2 g/kg of seeds 24 hours prior to soaking for sprouting. The treated seeds can be stored for 30 days without any loss in viability.

b) *Wet seed treatment*

The seeds should be treated with carbendazim at 2g / lit of water for 1 kg of seed and soaked for 2 hrs. Then after sprouting, seeds should be sown in the nursery bed. This wet seed treatment gives protection to the seedling disease such as blast. This method is better than dry seed treatment.

Selection of nursery area: 20 cents (800 m^2) of land should be selected near the water source for raising seedlings per hectare.

Application of organic manure: 1000 kg of FYM or compost should be applied to 20 cents nursery and spread the manure uniformly on dry unprepared soil.

Preparation of nursery

The nursery area should be flooded one or two days before ploughing to a depth of 2.5 cm and ploughed and brought into a puddled condition. Before the last puddling, 40 kg of DAP or straight fertilizers 16 kg of urea and 120 kg of super phosphate should be applied. Basal application of DAP is recommended when the seedlings are to be pulled out in 20-25 days. Application of DAP is to be done 10 days prior to pulling out. In clayey soil, where root snapping is a problem, DAP has to be applied @ 1 kg/cent 10 days after sowing. Three packets (600 g/ha) of Azospirillum culture are to be mixed with sufficient water wherein the seeds are soaked overnight before sowing in the nursery. After decanting the culture is poured into the nursery area itself. Seed treatment with *Pseudomonas fluorescens* is done for the management of rice blast. Three packets (600 g/ha) of *Pseudomonas fluorescens* peat culture should be added in water wherein the seeds are soaked overnight before sowing in the nursery bed. *P. fluorescens* can be mixed with Azospirillum culture as it is not inhibitory to Azospirillum. Keeping a thin film of water in the nursery, the sprouted seeds are sown uniformly on the seed beds.

Water management

Drain the water 18 to 24 hours after sowing and if there are pockets of water stagnation, drain it to the channel. Germination will be affected in places where there is water stagnation. Allow enough water to saturate the soil from 3 to 5 DAS. From the 5th day onwards, increase the quantity of water to a depth of 1.5 cm depending on the height of the seedlings. Afterwards, maintain 2.5 cm depth of water.

Weed management in nursery

Pre-emergence herbicides *viz*., butachlor 2.0 lit/ha or thiobencarb 2.0 lit/ha or pendimethalin 2.5 l/ha or anilophos 1.25 lit/ha may be applied on 8th DAS to control weeds in the lowland nursery. Safener mixed herbicide (Pretilachlor + fenclorim) can also be applied on 3 or 4 DAS to achieve early and effective weed controls.

Main field preparation

Puddling

Wet rice requires a well puddled soil. Ploughing the land with water is called puddling. The formation of impervious layer or pan formation is a result of repeated ploughing and puddling. It will prevent the downward movement of water in subsoil by percolation and loss of nutrients and also prevent the ground water upsurge thus bringing salts, toxic compounds or strongly deoxygenated water to the root zone.

Digging corners and maintenance of bunds

The corners of fields should be dug which are not covered by ploughing. 2.5 cm of soil from the top and side of the bunds should be cut off to remove the weeds along with their seeds and to destroy the eggs of insect pest by using spade. If rat burrows are noticed, pellets of 0.5 g or 0.6 g aluminium phosphide are applied in the existing holes. Plastering the bunds helps in checking weed growth and prevents harbouring of insect pests.

Steps in field preparation

The field is flooded one or two days before ploughing by allowing the water to soak the surface of the field. The water depth of 10 cm is maintained at the time of land preparation. The field is puddled with puddlers i.e., three puddling with puddlers or by use of cage wheel mounted on a tractor once or four ploughing with country plough with sufficient water. Perfect leveling is more important to avoid stagnation of water. 12.5 tonnes of FYM or compost per hectare should be applied. If the green manure crop is raised in the field itself it should be ploughed and incorporated.

List of green manure

Green manure	Scientific name	Biomass (t/ha)	N content (%)
Daincha	*S. aculeata*	10-15	3.0
Manila agathi	*S. roatrata*	25-30	3.5
Seemai agathi	*S. speciosa*	25	2.7
Sunnhemp	*Crotolaria juncea*	8-12	2.3

Application of inorganic fertilisers

Fertilizers should be applied based on soil test recommendations. If recommendations are not available, following blanket recommendation should be restored.

General recommendation	Fertilizers (kg/ha)		
	N	P_2O_5	K_2O
Short duration varieties	120	40	40
Medium duration varieties	150	50	60
Long duration varieties	150	50	80

Hybrids	150	60	60
Time of application	N	P_2O_5	K_2O
Basal	25%	100 %	100 %
1st Top dressing (AT)	25%	-	-
2nd top dressing (PI)	25%	-	-
3rd Top dressing (Heading)	25%	-	-

Bio-fertilizers application to rice

The commonly used bio-fertilizers for rice crop are: Azolla, Blue green algae, Azotobacter, Azospirillum and Phosphobacteria

Soil application Two kg of Azospirillum inoculant (10 pockets/ha) with 25 kg of powdered FYM and 25 kg of soil should be mixed and broadcasted in the mainfield at the time of transplanting.

Transplanting

Planting of rice seedlings in the puddled field is called transplanting. Water is allowed to the nursery to a depth of 10 cm to facilitate easy pulling out of seedlings and for washing the mud sticking to the roots. While pulling out seedlings, root damage should be avoided. Two seedlings are planted per hill at 15 x 10 cm spacing for short duration varieties and 20 x 10 cm spacing for medium duration varieties. Slurry is prepared with 5 pockets of azospirillum in 40 lit water and the root portion of the seedlings is dipped for 15-30 minutes in the suspension and transplanted.

Management of aged seedlings

Three to four seedlings per hill are planted deeper. When closer spacing is adopted, 25% extra N is added.

Gap filling and population management

Optimum plant population is maintained for achieving the targeted yield. (E.g.) 50 and 60 to 80 hills per m^2 is required for short and medium duration varieties. The gaps are filled within 7-10 days after planting.

Water management

Rice is a semi-aquatic plant. Its water requirement is many times more than other crops due to the water requirement for field preparation i.e., puddling (20 per cent) and percolation loss (43%). Moisture stress at critical stages leads to reduction in yield. Excess water or water stagnation causes salinity or alkalinity; weed problem and results in poor aeration and accumulation of toxic products.

Water requirement

Total water requirement includes water needed to raise seedlings, land preparation and to grow the crop from transplanting to harvest. Water requirement is 900, 1200 and 1500 mm for short, medium and long duration varieties respectively. Daily consumptive use of water is 6 to 10 mm/day.

Water requirement for various operations is:

Nursery preparation	:	150 to 200 mm
Raising of seedlings	:	250 to 400 mm
Land preparation	:	300 mm
Field irrigation	:	1000 mm

Critical stages for irrigating rice are primordial initiation, booting, heading and flowering.

Common water management practices:

A. Continuous submergence or flooding static

a) Shallow (2.5 cm water depth)

b) Medium (2.5 cm to 7.5 cm water depth)

c) Deep (15 cm water depth)

B. Continuous flooding (submergence)

In some terraces in the valley, irrigation water is abundant due to surplus supply of water coming from the mountain. In these areas this system of water management is practiced.

C. Rotational irrigation or intermittent irrigation

It is practiced in areas which are often subjected to water shortage or where no abundant supply of water for continuous irrigation exists.

Drainage of water in rice

Drainage of water at maximum tillering stage stimulates vigorous growth of roots and checks the development of unproductive tillers. Irrigation is stopped 10 – 15 days before harvest. If irrigation is not stopped at appropriate time it leads to increase in crop duration and shattering of grains.

Weed management

Pre emergence herbicide: Any one of the herbicides like butachlor 1.25 kg/ha, thiobencarb 1.25 kg/ha, pendimethalin 1.0 kg/ha, anilofos 0.4 kg/ha or pretilachlor 0.75 kg/ha is applied on 3 days after planting using 50 kg sand. Water depth is maintained at 2.5 cm at the time of herbicide application.

Post emergence herbicide:

Fernoxone (2,4-D sodium salt) : 1.00 kg/ha

Propanil (Stam F-34) : 1.25 kg/ha

For controlling sedges and broad-leaved weeds, 2,4-D sodium salt is sprayed. To control both grassy weeds and broad-leaved weeds a combination of propanil (0.6 lit/ha) and fernoxone (1.0 kg/ha) is applied.

Harvesting

Field should be drained 7 to 10 days before the expected date of harvest or when the upper grains in most of the tillers are in the hard dough stage and turning from green to yellowish. This operation hastens the maturity of the crop. The rice plants are cut manually with serrate edged sickles. When the land surface is fairly firm, short stubbles alone are left in the field. The combine is more useful to complete the operation in time especially when labour shortage is noticed. The area coverage is also more compared to manual harvest.

Post-harvest operations

Threshing

The ideal moisture content for threshing either manually or mechanically is 15- 18%. Threshing can be done by manual labourers, animals and machines. Electric or diesel engine operated power threshers are more efficient. Winnowing is also completed. In power threshers entire produce is obtained as a cleaned paddy. Straw is also separated without any damage.

Drying

Seed should be dried to bring the moisture content to 12 – 14%. Sun drying is a traditional method of drying of grains adopted by the farmers. Mechanical drying is the process of utilizing mechanical means for drying of grains by ventilating natural or heated air through grain mass to accomplish removal of moisture from it.

The average outturn from field paddy is 66%. Rice processing operations include threshing, parboiling, milling and grading.

Storage

The recommended storable moisture content for paddy is 13%. Bag and bulk storage are the two basic methods of storage.

Systems of rice cultivation

1. Lowland:	a) Tank	b) Canal
2. Upland:	a) Rainfed (Dry)	b) Bunded (Semi-dry)

1. Lowland Cultivation

It is irrigated rice, which occupies one half of the global rice. Nearly 75% of the world rice production is derived from humid and sub humid tropics under irrigated conditions. In this system, rice is grown under wet conditions under submergence. There are two methods of cultivation.

a) Direct sowing: Seeds are sown straight to the main field either by broadcasting or row sowing. This method is practiced in some areas where water is not sufficient.

b) Transplanting: Seedlings are raised on the seed bed before planting in the main field. In command areas, transplanting method of rice is being practiced by the farmers. Transplanting by manual labour is time consuming and difficult due to the non-availability of labour. Mechanical transplanters enhance the efficiency of the transplanting process.

2. Upland cultivation

Upland rice is common in West Africa, Latin America and Asia. In Southern America it is seen in sloppy areas.

a) Dry cultivation: It is practiced in all rice growing states in India but are mainly confined to tracts which get rain either by the South-West or North-East monsoon or both and do not have adequate irrigation facilities. Upland rainfed rice cultivation is commonly found in hilly areas and coastal belts.

b) Semi-dry cultivation: In this system, seeds are sown broadcast as a dry crop and when more water is available after the receipt of monsoon, it is treated as wet crop. In South-West Monsoon it is practiced in Kanyakumari, (*Kannipoo paruvam*) and in North-East Monsoon it is practiced in Chengalpattu, Pudukottai, Ramanathapuram, Sivagangai and Virudhunagar districts.

3. Deep water rice

Flooded rice occupies 10 to 12 million hectares. In Bihar, West Bengal and Assam deep water rice is grown in 4-5 meters depth of water. The seed is broadcasted with the pre-monsoon showers. As the level of water in the field rises the varieties keep pace with the rise of water. The ears of the crops are harvested with the help of boats since the level of the water in the field remains high till harvest. In parts of Uttar Pradesh, the seeds are broadcasted by harrowing or laddering at 80 to 120 kg/ha.

Direct seeding under wet land condition

Direct seeding is a recommended practice for wet land condition. The seeds are broadcasted in the main field after thorough puddling and land leveling as in the nursery.

Advantage in direct seeding

It reduces the main field duration by 15 days. It lowers the water requirement and withstands early stress. It gives equal or better yield compared to planted rice.

Seed rate: 60 kg/ha.

Sowing

Broadcasting / drum seeding method is followed. Broadcasting is the conventional method of sowing where population is not assured. Drum seeding is proved to be more effective method due to the assured plant population and it is easy to adopt. With two 'A' type labours one acre sowing can be completed. Modified drum seeder is used for simultaneous sowing of green manure and rice seeds in alternate rows.

Fertilizer

- 120:40:40 kg NPK/ha for SDV
- 150:50:50 kg NPK/ha for MDV

Nitrogen is applied in the form of urea on 15, 25 and 45 DAS. Phosphorus and potassium as super phosphate and muriate of potash respectively are applied basally.

Precautions in wet seeding:

a) Thorough main field preparation and land leveling

b) Avoid water stagnation especially during early stage of growth

c) Early weed control

d) Avoid basal application of N fertilizer as this promotes more weed growth

e) Uniform water application at the early stage

Weed control

Any one of the pre-emergence herbicides viz., butachlor 1.25 kg, thiobencarb 1.25 kg, pendimethalin 1.00 kg per ha is applied as sand mix on 8 DAS. But early weed control is more desirable which is achieved by using pretilachlor + safener @ 0.4 kg/ha on 3 or 4 DAS.

Difference between transplanted and direct seeded rice

S.No.	Transplanted	Direct seeded
1.	More duration	Less duration by 10-15 days
2.	Less weed problem	More weed problem
3.	More water requirement	Less water requirement
4.	More yield	Less yield
5.	Cost of cultivation more	Less cost of cultivation
6.	Nursery is required	No nursery is required

Rainfed rice cultivation

Conditions for growing rainfed rice:

a) Rainfall should be uniform and well distributed during the season

b) The average annual rainfall should exceed 850 mm.

Suitable places: Chengalpattu, Pudukottai, Ramanathapuram, Virudhunagar, Sivagangai and Kanyakumari Districts of Tamil Nadu.

Season

June – July to Sep – Oct and Oct – Nov to Jan – Feb.

Field preparation

The field has to be prepared under dry condition to a very fine tilth, taking advantage of summer rains and pre-monsoon showers. Apply farm yard manure @ 12.5 t/ha. Spread the manure evenly on the dry soil and incorporate.

Traditional varieties

The important traditional varieties are Chitraikar, Chandikar, Kuruvaikalanjium, Kulipudichan, Vellaikar, Aryanmanavari, Kathanur, Vadakathikar, Poonagar, Kallurandaikar, Sadai-samba, Chithariyan 110 and Maruthakar.

Varieties

The important varieties are ASD 1, MDU 1, MDU 2, MDU 3, PKM 1, TPS 1, TPS 2, TKM 1, TKM 2, TKM 9 and Anna 4, ADT 36, TKM 10, TKM (R) 12, PMK (R) 3, BPT 5204, RMD (R) 1, Nootripathu, Norungan.

Seeds and sowing

Seed rate required is 75 to 100 kg of dry seed per ha for any recommended variety. Treat the seeds as adopted for wet paddy. After sowing, sheep penning for a week helps to compact the soil and even sprouting. Broadcast the seeds and cover immediately by harrows or sow by local "gorrus" for line sowing.

Seed hardening

The seeds should be soaked in water for 10 hours and then in 1% potassium chloride solution (dissolve 400 g of muriate of potash in 10 lit of water to prepare 1% solution) and shade dried. This treatment gives healthy seedlings in rainfed condition and help to produce deeper roots thereby the crops get the capacity to withstand drought.

After cultivation

Thinning and gap filling should be done at 10-12 days after sowing taking advantage of the immediate rain.

Nutrient management

Usually, no basal dressing of nitrogen is done for dry direct seeded rice. If rains are favourable 50 kg of N/ha can be top dressed at 45-60 days after sowing taking advantage of moisture through rain.

Nitrogen management

Application of less amounts of nitrogen fertilizer at the time of sowing discourages weed growth. Most of the nitrogen is applied during the period of tillering and panicle initiation. In light soils at least 2-3 split application of N are needed. The field should be kept free from weeds for maximum utilization of nitrogen by the crops.

Weed management

Between 15 and 20 days, first weeding should be done. Second weeding can be done at 45 days after first weeding. Use thiobencarb @ 2.5 lit/ha or pendimethalin @ 2.75 kg/ha at 8 DAS if adequate moisture is available. Spraying insecticides and fungicides are need based. Working wooden planks on alternate days upto 12th day is practiced during first crop in Kanyakumari district. This is done for breaking clods, compaction and for weed control. In Chengalput district harrows are used to stir the soil in between the lines.

Harvesting

Harvesting practices are same as that of wet rice cultivation.

Semi dry rice cultivation

Field preparation, choice of varieties and seeding practices are same as that of dry rice cultivation till the receipt of monsoon rains. After the receipt of North-East monsoon rains, irrigation tanks get filled up. The tank water is used for irrigating the rice crop and the crop is treated as wet land rice from 30 to 45 days after sowing. Until flooding, this crop is maintained as a dry crop.

Questions

I. Choose the best from the choices given

1. The method in which seedlings are raised without contact with soil which are ready for transplanting on 12th day

   a) Wet nursery b) Dry nursery

   c) Dapog nursery d) None

2. Water requirement for medium duration rice varieties is

   a) 900 mm b) 1200 mm

   c) 1500 mm d) 1800 mm

3. Seed rate for short duration rice varieties is

   a) 20 kg /ha b) 30 kg /ha

   c) 40 kg /ha d) 60 kg /ha

4. Seedlings of medium duration rice varieties are transplanted at

   a) 21-24 days b) 25-30 days

   c) 30-35 days d) 36-40 days

5. Planting of rice seedlings in the puddled field is called
   a) Sowing b) Transplanting
   c) Seedling throwing d) Dibbling

II. Fill in the blanks

1. The green manure generally recommended for rice is _________
2. The optimum population for short duration rice is ____________
3. The fertilizer requirement for medium duration rice is __________ kg NPK / ha.
4. The recommended storable moisture content for paddy is ____ %
5. The seed rate for direct seeded rice is __________ kg / ha.

III. Write short notes

1. Weed management in Rice
2. Rice seasons of India
3. Dapog method of nursery
4. Puddling
5. Management of aged seedlings

5

Special Type of Rice Cultivation SRI and Hybrid Rice Cultivation

System of rice intensification (SRI)

System of Rice Intensification (SRI) was first developed in Madagascar during the 1980s SRI uses less inputs. It uses less seed, water, chemical fertilizers and pesticides but uses more organic manures. Rice grown with SRI technology has large root volume, profuse and strong tillers with big panicles, more and well-filled spikelets with higher grain weight.

SRI is initially labour intensive and this method needs 50% more-man days for transplanting and weeding. It mobilizes labour to work for profit. It offers an alternative to resource poor, who puts in their family labour. Once skills are learnt and implements are used, the labour costs will be lesser than the present-day rice cultivation.

SRI encourages rice plant to grow healthy with large root volume, profuse and strong tillers, non-lodging, long panicle, more and well filled spikelet's, more panicle weight and higher grain weight and resists insects because it allows rice to grow naturally

Tillering is greatly increased. Thirty tillers per plant are fairly easy to achieve. Fifty tillers per plant are quite attainable. With really good use of SRI, individual plants can have 100 fertile tillers or even more.

The six basic principles of SRI are

1. Use of young seedlings for transplanting
2. Careful transplanting
3. Planting at wider spacing
4. Weed control
5. Water Management
6. Organic manures

1. **Early transplanting**

   Seedling 8-12 days old, when the plant has only **two small leaves,** before fourth phyllochron

2. **Careful transplanting**

   Minimize trauma in transplanting. Remove the seedling from nursery with the seed, soil and roots carefully and place it in the field without plunging too deep into soil.

3. **Wide spacing**

   Single seedling is planted in a square pattern at a spacing of 25 x 25 cm.

4. **Weeding and aeration**

   Weeding is done by Cono weeder or a simple mechanical "rotating hoe" that churns up soil; 2 weedings required, with 4 recommended before panicle initiation; first weeding 10 days after transplanting.

5. **Water Management**

   Regular water applications to keep the soil moist but not saturated, with intermittent dryings, alternating aerobic and anaerobic soil conditions.

6. **Compost / FYM applied instead of or in addition to chemical fertilizer**

   Ten tonnes/ha has to be applied In SRI Cultivation 8 to 12 days old seedlings are planted. So, root system grows well and gives 30 to 50 tillers. When all the 6 management practices are followed then 50 to 100 tillers are produced per plant and high yields can be realized.

Season

Dry season with assured irrigation is more suitable. Difficulty in crop establishment may be seen in areas with heavy downpour (North east monsoon periods of Tamil Nadu)

Varieties: Hybrids and varieties with heavy tillering.

Generally, the long-duration varieties of rice perform better with wider spacing, which is the recommended practice under SRI, than the short-duration varieties of rice. Latif et al. (2005) reported that long-duration varieties and medium duration also more suitable for SRI practices, as they led to a higher grain yield.

ADT 47, ADTRH 1, Swarna, CO (R) 48, CO RH 3, White ponni, Mugad Sugandhi, GEB 24, Jeeragasamba, KRH 1, Ajay, Paiyur 1, Savitri, CRHR 7, Khandagiri

NURSERY

Seed rate: 7- 8 kg for single seedling/hill.

Preparation of nursery area

Prepare 100 m^2 nursery to plant 1 ha area. Select a leveled area near the water source. Spread a plastic sheet or used polythene gunny bags on the shallow raised bed to prevent roots from growing deep into soil.

Preparation of soil mixture: Four (4) m^3 of soil mix is needed for each 100 m^2 of nursery. Mix 70% soil + 20% well-decomposed pressmud / bio-gas slurry / FYM + 10% rice hull. Incorporate in the soil mixture 1.5 kg of powdered di-ammonium phosphate or 2 kg 17-17-17 NPK fertilizer.

Filling in soil mixture: A wooden frame of 0.5 m long, 1 m wide and 4 cm deep divided into 4 equal segments is placed on the plastic sheet or banana leaves. The frame is filled almost to the top with the soil mixture.

Pre-germinating the seeds 2 days before sowing: The seeds are soaked for 24 hr, water drained and the soaked seeds are incubated for 24 hr. Sowing is done when the seeds sprout and radical (seed root) grows to 2-3 mm long.

Sowing: The pre-germinated seeds weighing 90-100 g/m^2 (100g dry seed may weigh 130g after sprouting) may be sown uniformly and covered with dry soil to a thickness of 5mm. Water is sprinkled immediately using rose can to soak the bed and the wooden frame is removed and the process is continued until the required area is completed.

Watering: the nursery is watered with rose-can as and when needed (twice or thrice a day) to keep the soil moist. The nursery should be protected from heavy rains for the first 5 DAS. At 6 DAS, a thin film of water is maintained all around the seedling mats. The water is drained 2 days before removing the seedling mats for transplanting.

Spraying fertilizer solution (optional): If seedling growth is slow, 0.5% urea + 0.5% zinc sulfate solution can be sprinkled at 8-10 DAS.

Lifting seedling mats: Seedlings reach sufficient height for planting at 15 days. The seedling mats are lifted and transported to main field.

Main field preparation

Land preparation is not different from regular irrigated rice cultivation. Levelling should be done carefully so that water can be applied very evenly. At every 3m distance form a canal to facilitate drainage. With the help of a marker draw lines both way at 25 x 25cm apart and transplant at the intersection

Transplanting

Single seedling of 15 days old is used for transplanting. Square planting is adopted at a spacing of 25 x 25 cm. The gaps are filled between 7 and 10 DAT. Transplant within 30 minutes of pulling out of seedlings.

Irrigation management

Irrigation only to moist the soil in the early period of 10 days. Restoring irrigation to a maximum depth of 2.5cm after development of hairline cracks in the soil until panicle initiation. Irrigation depth is increased to 5.0cm after panicle initiation one day after disappearance of ponded water

Weed management

Rotary weeder / Cono weeder is used for weeding. Moving the weeder with forward and backward motion to bury the weeds and as well to aerate the soil at 7-10 days interval from 10-15 days after planting on either direction of the rows and column.

Nutrient management

As per transplanted rice. Use of Leaf colour chart (LCC) has more advantage in N management.

N management through LCC

Time of application is decided by LCC score. Observations are taken from 14 DAT in transplanted rice or 21 DAS in direct seeded rice. The observations are repeated at weekly intervals up to heading. The leaf colour is observed in the fully opened third leaf from the top as index leaf. The leaf colour is matched with the colours in the chart during morning hours (8-10 am). Observations are taken in 10 places. LCC critical value is 3.0 in low N response cultures like 'White ponni' and 4.0 in other cultivars and hybrids. When 6/10 observations show less than the critical colour value, N can be applied @ 35kg N/ha in dry season and 30kg N/ha in wet season per application. Green manure and farm yard manure application will enhance the growth and yield of rice in this system approach.

Other package of practices as recommended to transplanted rice

Benefits of SRI

Higher yields of both grain and straw can be obtained. The duration is reduced by 10 days. Water requirement is less. Chaffy grain per cent is less. Grain weight increased without change in grain size. Higher head rice recovery. Withstand cyclonic gales. Soil health is improved through biological activity.

Comparison of SRI with conventional method:

S.No	Conventional Method	SRI Method
1.	20-25 kg seed is used per acre	2 Kg seed is sufficient for one acre
2.	25 to 30 day old seedlings are transplanted.	Only 8-12 day old seedlings transplanted.
3.	33 hills are planted per Sq. m	16 hills are planted per Sq. m or less
4.	3 or more plants planted in clumps	Only one plant is planted per hill.
5.	Application of NPK fertilizers recommended	Application of organic manures, only basal dose of fertilizers, no top dressing
6.	Continuous flooding	Moist condition
7.	Herbicide is used for weed management	Cono weeder is used for weed management

Hybrid rice

Hybrid rice has yield advantage of 15 per cent over the best conventional varieties. Short (CoRH 1 and ADTRH 1) and medium (CoRH 2) duration hybrids are chosen for *kharif* and *rabi* season, respectively. Seed requirement for hybrid is 20 kg/ha. Sparse sowing in nursery @ 2 kg seeds per cent is recommended. This favours tillering and makes it possible to transplant a single hybrid seedling per hill with 3-4 tillers.

Twenty-five days old seedlings are transplanted at 20 x 10 or 20 x 15 cm spacing. The optimum plant stand is about 40-50 hills per m^2. Farm yard manure or green manure @ 6.25 t/ha is applied. Azospirillum is recommended as seed treatment (600 g), seedling dipping (1000 g) and soil application (2000 g). Higher grain yield is obtained with 150 to 225 kg N/ha. Phosphorus and potassium are applied each at 50 kg per ha. Entire quantity of phosphorus is applied as basal dose during last puddling. Nitrogen and potassium are applied in four equal splits *viz*., active tillering, panicle initiation, flowering and heading stages. The major constraint in hybrid rice is poor seed setting.

Cropping system in wetland

Rice-wheat system is practiced under irrigated condition in Punjab, Haryana and Western UP and as rainfed system in Bihar, Orissa, Wet Bengal and Assam. Rice-winter maize/gram is the next important system. Rice-jute/potato is also followed in West Bengal. Rice-sugarcane system is adopted in Maharashtra.

In Tamil Nadu, rice-rice-rice, rice-rice-pulse/gingelly/cotton, rice-rice-groundnut, maize-rice-pearl millet and rice-black gram-gingelly-green gram and rice-rice-

rice-green manure are some of the rice-based cropping systems followed in double crop wetland. In Cauvery delta zone, rice-groundnut-soybean, rice-cotton-ragi, rice-rice-ash gourd and rice-soybean-maize systems are followed. In single crop wetland rice-pulse/ gingelly/cotton is adopted.

Aerobic rice

Aerobic cultivation system is the method of cultivation, where the rice crop is established by direct seeding (dry or water-soaked seed) in non-puddle field and un-flooded field condition. It is the most promising approaches for saving water and labour. It is grown like an upland crop in soil that is unpuddle, non-flooded or saturated. The soil is therefore 'aerobic' or with oxygen through the growing season. In Tamil Nadu the variety suitable for aerobic rice is PMK 2.

Golden rice

Half of the world population's main food source is rice. The main concern about rice is that it has insufficient concentrations of vitamin A. It has been suggested that rice could be fortified to reduce the level of nutritional vitamin A deficiencies. Golden rice was originally created by Dr. Ingo Potrykus and his team in Zurich, Switzerland. This genetically modified rice is capable of producing beta-carotene in the endosperm (grain) which is a pre cursor for vitamin A production.

Questions

I. Choose the best from the choices given

1. In SRI method of Rice cultivation, recommended age of seedlings for transplanting is

   a) 30 days b) 20 days

   c) 15 days d) 8-12 days

2. Cono weeder is used under

   a) Direct planting system in rice

   b) Conventional method

   c) SRI method

   d) Direct seeding method

3. The recommended spacing under SRI method is

   a) 20 x 15 cm b) 15 x 10 cm

   c) 25 x 25 cm d) 20 x 15 cm

4. Seed requirement for hybrid rice is

   a) 10 kg / ha b) 15 kg / ha

   c) 20 kg / ha d) 25 kg / ha

5. Yield advantage of hybrid rice over conventional rice is

   a) 10 % b) 15 %

   c) 25 % d) 30 %

II. Fill in the blanks

1. SRI was first developed in __________.
2. The age of seedling for SRI type of rice cultivation is _________
3. Nursery area required under SRI to transplant one ha is _______.
4. The seed rate to plant one acre under SRI is __________.
5. Number of plants planted per hill under SRI is _________.

III. Write short notes

1. Basic principles of SRI
2. Benefits of SRI
3. Rice based cropping system
4. Rainfed rice
5. Hybrid rice

6

Maize - Origin, Geographic Distribution, Economic Importance Soil and Climatic Requirement Varieties, Cultural Practices and Yield

Maize (*Zea mays*)

Origin

The primary centre of origin of maize is Central America and Mexico. During seventeenth century, it was introduced in India by the East India Company.

Distribution

Maize is one of the world's leading crops cultivated over an area of about 194 million hectares with a production of about 1147 million tonnes of grain (2020). Among the maize growing countries, USA has the largest area followed by China, Brazil, Argentina, Ukraine and India. In respect of production also USA stands first followed by China. Regarding average yield per hectare USA ranks first followed by China. The average maize yield in India is only 3065 kg per hectare which is much lower than most of the maize growing countries of the world.

In India it is grown over an area of 9.57 million hectares with total production of about 28.75 million tonnes. Rajasthan, Karnataka, Uttar Pradesh, Bihar, Madhya Pradesh and Andhra Pradesh are the leading states growing maize on large scale. Though the maximum area is in Madhya Pradesh, Karnataka ranks first in total production.

Economic importance

Maize is a staple human food and feed for livestock. It is used for fermentation and many industrial uses. It is rich in starch (65%) and used for industrial starch production mainly sweeteners. Wet milling industries also produce various modified maize starch for paper lamination, textile wrap, sizing and laundry finishing. Whereas, dry milled products are animal feed, brewing, breakfast cereals, other food. In India, dry milling is the predominant process for both flour and animal feed, fermentation and distilling industries and composite flours. In the new millennium it is an alternate crop to rice and wheat origin. Among the C4 plants, maize gives higher yield and hence it is called "Queen of Cereals".

Nutritive value

Maize protein 'Zein' is deficient in tryptophane and lysine, the two essential amino acids. Maize is low in calcium and fairly high in phosphorus.

Classification

Maize (*Zea mays*) is an annual plant which belongs to family Gramineae and Genus *Zea*. It is divided into seven groups. The classification is based largely on the character of the kernels.

1. ***Zea mays indurata*** **or 'Flint corn':** The endosperm in this type of maize kernel is soft and starchy in the centre and completely enclosed by a very hard outer layer. The kernels are usually round but are sometimes short and flat. Colour may be white or yellow. This is the type most commonly cultivated in India.
2. ***Zea mays indentata*** **or 'Dent corn':** In this type of maize kernels have both hard and soft starches. The hard starch extends on the sides, and the soft starch is in the centre and extends to the top of the kernels. In the drying and shrinking of the soft starch, various forms and degrees of indentation result. This is the common type of maize grown in USA
3. ***Zea mays everta*** **or 'Pop corn':** It possesses exceptional popping qualities. Size of the kernels is small but the endosperm is hard. When they are heated the pressure built up within the kernel suddenly results in an explosion and the grain is turned inside out.
4. ***Zea mays saccharata*** **or 'Sweet corn':** Kernels possess a considerable amount of sugar which absorbs water, making the cells turgid, on drying these cells collapse, making the grains shrivelled or wrinkled. It has sweeter taste than other corns.
5. ***Zea mays amylacea*** **or 'Soft corn':** It possesses a soft endosperm. Kernels are soft and of all colours, but white and blue are the most common. They are like flint kernels in shape.
6. ***Zea mays tunicata*** **or 'Pod corn':** The pod corns are characterized by having each kernel enclosed within a pod or husk. It is a primitive type of corn and hence of no importance.

7. ***Zea mays Ceratina Kulesh* or 'Waxy corn':** The endosperm of the kernel when cut or broken gives a waxy appearance. It produces the starch similar to tapioca starch for making adhesive for articles.

Inflorescence: Maize is a monoecious plant having both male and female inflorescences on the same plant. Male flowers are called 'tassel'. Tassel is born at the top of the stem and female flowers are called 'silk' and it is borne inside the young cobs.

Kernel or caryopsis: Maize kernel is one seeded fruit or caryopsis. Seed enclosed within the pericarp consists of the embryo, endosperm and remnants of seed coat and nucellus.

Climate

Maize is a warm weather plant. It grows from sea level to 3000-meter altitudes. It can be grown under diverse conditions. It is grown in many parts of the country throughout the year. *Kharif* (monsoon) season is the main growing season in northern India. In the south, however, maize may be sown any time from April to October, as the climate is warm even in the winter. Maize requires considerable moisture and warmth from germination to flowering. The most suitable temperature for germination is 21°C and for growth 32°C. Extremely high temperature and low humidity during flowering damage the foliage, desiccates the pollen and interferes with proper pollination, resulting in poor grain formation. It can be successfully grown in the areas receiving an annual rainfall of 600 mm well distributed throughout its growth period. Maize is very sensitive to stagnant water, particularly during its early stages of growth.

Soil

Maize is the best adapted crop to well drained sandy loam to silty loam soils. Water stagnation is extremely harmful to the crop; therefore, proper drainage is a must for the success of the crop especially during *Kharif* season. Maize will not thrive on heavy clays, especially low lands. It can be grown successfully in soils whose pH ranges from 5.5 to 7.5. The alluvial soils of Uttar Pradesh, Bihar and Punjab are very suitable for growing maize crop.

Varieties

The promising maize hybrid and composite varieties for various maturity groups for Tamil Nadu are Deccan-103, Ganga-11, COH-2, COH-3, CO HM 5, CO 6, Deccan-107, KH-510, Deccan-109, Prakash, Vivek-9, Pusa hybrid-1, Pusa hybrid-2, Ganga 5, TNAU maize hybrid CO 6, CO(H) 8, Baby corn CO (Bc) 1, Pusa Vivek QPM improved (Provitamin-A, lysine & tryptophan rich hybrid), Pusa HM4-HM8 & HM9 improved (lysine and tryptophan rich hybrid)

Other varieties

Vijay, Kisan, NLD White, Naveen, Diara, Ganga Safed-2, Ganga -101, Deccan 103, DHM 103, Ganga II, PRO 311 (4640) Bio 9681, SSF 9374 COH2, COH3, JK 2482, Y 1402 K (3058), Trishulata, KH 5991, 5981, Dhawal, Pusa Hybrid -1, Pusa Hybrid -2, Central Maize VL (Sweet corn), COH (M) 7, 8,9 &10, Vivek Maize Hybrid 47, 51 & 53, BH-1620 (DHM113), HM-8 (HKH1188), BPCH-6 (Popcorn), Vivek Hybrid 27 (Babycorn), Vivek QPM 5, 7 & 9.

Field preparation

Maize kernels need a seedbed which is friable, well aerated, moist and weed free to provide better contact between the seed and the soil. There is no need of preparing an extremely fine seedbed. The first ploughing should be done with soil inverting plough so that at least 20-25 cm deep soil may become loose. It should be followed by two to three harrowings or three to four intercrossing ploughings with local plough. Planking should be done after each ploughing. A properly levelled and uniformly graded field is required for good water management.

Seed rate and spacing

Maize requires spacing of 60 to 75 cm from row to row and 20 to 25 cm from plant to plant for higher grain yield. For higher altitude, the recommended spacing is 45 cm x 20 to 25 cm and for mid to low altitude, the spacing is 60 cm x 25 cm. The recommended seed rate is 20 to 25 kg/ha for grain crop and 30 to 40 kg/ha for fodder crop. For hybrids the seed rate is 20 kg / ha.

Seed treatment

Maize seeds should be treated with fungicides like captan or thiram or emisan @ 2 to 3 g/kg of seeds to protect the crop from seed borne diseases viz., leaf blight in maize.

Time of sowing

Maize crop can be grown throughout the year mainly due to the availability of thermo and photo-nonsensitive maize varieties. Usually, three crops of maize viz., *kharif, rabi* and *zaid* are grown in the country and sowing time differs accordingly.

Kharif crop: It is observed that after outburst of monsoon the rains impose a serious problem in sowing of the seed, therefore, it is always better to sow the crop at least a fortnight before rain starts so that the seedling is well established before they get torrential rains. Early sowing also reduces the weed competition, the crop matures earlier and the field becomes free for an early *rabi* season crop. *Kharif* sowing is done at different times in different regions:

Rabi crop: This crop is popular in Bihar, parts of U.P., Andhra Pradesh, Tamil Nadu, Karnataka and Maharashtra. The optimum date of sowing is end of October to mid of November. The crop is raised under assured irrigation.

Spring Sowing: This crop is grown in Bihar, Maharashtra, and tarai belt of U.P. and the most optimum time for crop planting is late January to mid-February. The crop is grown irrigated.

Method of Sowing

The crop is sown in for different methods *viz.,*

1. **Broadcasting**: This method is still common among some backward farmers but it is unscientific, requires more seed rate, imposes problem in weeding etc. and hence this should be discouraged.
2. **Drilling Method**: In this method seed drills are used for sowing the seeds. Sometimes under high moisture condition of the soil it becomes difficult to run the seed drill because the sprouts are chocked with the wet soil and black rows are observed. The method is well suited for sandy soils and is very quick.
3. **Dibbling Method**: This method is time consuming as the seeds are placed by the help of a stick manually at required distance in the rows marked at 75 cm apart. It needs less seed rate and is the best method.
5. **Sowing the seeds behind the plough**: This method is very common in almost all the maize growing areas of the country. In this case shallow furrows are opened by the help of a country plough and seeds are dropped in them and planking is done

Intercultural operation

Thinning should be done keeping one plant per hill within 15 days after germination. First intercultural operation can be done within 2 to 3 weeks of germination to destroy weeds. A second hoeing and earthing up at 6-7 weeks stage is much useful practice.

Nutrient management

Manures and fertilizers both play important role in the maize cultivation. A liberal quantity of bulky manures should be applied in the field if available. Add 10 to 15 tonnes of farm yard manure or compost before sowing. The application of organic matter to the soil ensures good tilth and improves water-holding capacity.

Hybrids and composite varieties of maize exhibit their full yield potential only when supplied with adequate quantities of nutrients at proper time. As a general recommendation, one could apply 120 kg N, 60 kg P_2O_5 and 40 kg K_2O per hectare for hybrids and 80 kg N and 30 kg P_2O_5 and 20 kg K_2O per hectare for composites. In Tamil Nadu, the recommended dose of fertilizers are 250:75:75 N. P_2O_5 and K_2O /ha for hybrid. The entire P, ¼ N and ½ K is applied as basal and the remaining N is applied in two splits (½ at 25 DAS, ¼ at 45 DAS) and the remaining ½ K is applied at 45 DAS. TNAU micronutrient mixture @ 30 kg/ha is applied as enriched FYM.

Water management

Maize is very susceptible both to excess water and moisture stress. Water should not be allowed to stand in the maize field at any stage of its growth. Water stagnation even for six hours continuously and sufficiently damage the crop. Maize can tolerate heavy rains, provided water does not stand in the field for long periods. Therefore, excess water should be drained away by making a drain of adequate capacity at the lower end of the field. A good crop of maize requires about 460 to 600 mm of water during its life cycle. Maize plants should not be allowed to wilt due to water shortage at any stage of the life cycle. Tasseling to silking stage is critical. At this stage water shortage even for 2 days can reduce maize yields by about 20 per cent. The same for 6-8 days can pull down the yield by 50 per cent. The crop should be irrigated whenever it is needed. Drip Fertigation may also be adopted.

Weed management

The maize crop kept weed free for 30 to 45 days after planting is almost similar in yield as that kept weed free for entire crop season. Two to three manual weedings would be needed for this purpose. Generally, hand hoe and spades are used for weed control in maize. An effective way to control weeds is the use of pre-emergence herbicides *viz.,* Atrazine at 0.5 kg a.i./ha in light soils and 0.75 to 1.0 kg a.i./ha in heavy soils. If pulse crop is to be raised as intercrop, Atrazine should not be used. Pendimethalin@0.75 kg/ha has to be sprayed as pre emergence on 3-5 DAS.

Thinning and gap filling

Leave one healthy plant and remove other on 12 – 15 DAS. Gap filling may be done in non-germinated places.

Cropping system

In the irrigated areas many rotations of crops involving maize are feasible. Crops like wheat, potato, toria, sugarcane, chick pea, berseem, lucerne, barley and oats etc., can be grown after harvest of maize. Some of the most important cropping systems are given below.

1.	Maize-Potato	1 year
2.	Maize-Wheat	1 year
3.	Maize-Potato-Wheat	1 year
4.	Maize-Berseem (Egyptian clover)	1 year
5.	Maize-Wheat-Sugarcane	2 years
6.	Maize + Cowpea	1 year
7.	Maize + Blackgram	1 year
8.	Maize + Green gram	1 year

Crops like soybean, blackgram, moong, cowpea etc., are grown mixed with maize. These legume crops are grown in the space between two rows of maize. In Bihar mixed cropping of groundnut with maize is quite profitable. In certain areas maize with pigeon pea is also grown.

Harvesting and threshing

Maize crop is harvested when husk has turned yellow and grains are hard enough having less than 30 per cent moisture. Waiting for stalks and leaves to dry is not necessary because they remain green in most of the hybrids and composites. The husk should be removed from the cobs and then dried in sun for seven to eight days. Thereafter grains are removed either by beating the cobs by sticks or with the help of maize shellers.

Yield

The average yield of maize under improved cultivation practices is 5-6 t/ha in case of hybrids and 4.5 - 5.0 t/ha in case of composites under irrigated conditions. In case of rainfed crop yield levels are about 2.0 -2.5 t/ha for hybrids and 1.5 - 2.0 t/ha for composites.

Questions

I. Choose the best from the choices given

1. The water requirement of maize is
   - a) 50 mm b) 500 mm
   - c) 1000 mm d) 1500 mm
2. Soft corn is also known as
   - a) Flour corn b) Flint corn
   - c) Pod corn d) Dent corn
3. The crop called as queen of cereals is
   - a) Rice b) Maize
   - c) Wheat d) Sorghum
4. Female flower in maize is called
   - a) Silk b) Panicle
   - c) Tassel d) Cob
5. Most suitable temperature for growth of maize is
   - a) 30-35°C b) 10-15°C
   - c) 5-10°C d) 15-20°C

II. Fill in the blanks

1. The herbicide commonly used for maize is ______________.
2. Water requirement of maize is____________.
3. The best method of sowing maize is ___________.
4. In India mostly __________ type of maize is grown.
5. Optimum population/ha of maize for achieving higher yield is ________.

III. Write short notes

1. Uses of maize
2. Maize hybrids
3. Pop Corn
4. Weed management in maize
5. Intercropping in maize

7

Sorghum and Pearl Millet-Origin, Geographic Distribution, Economic Importance, Soil and Climatic Requirement, Varieties, Cultural Practices and Yield

Sorghum *(Sorghum bicolor* L. Moench)

Origin

Sorghum is probably originated in Africa. Because of the great diversity of types grown in East Central Africa, in and around Ethiopia and Sudan, it seems probable that it originated in that area (De Condole, 1884).

Distribution

It is grown in 98 countries of Africa, Asia, Oceania, and the Americas. Nigeria, India, USA, Mexico, Sudan, China and Argentina are the major producers. Other sorghum producing countries are Mauritania, Gambia, Mali, Burkina Faso, Ghana, Niger, Somalia and Yemen, Chad, Sudan, Tanzania and Mozambique. Among the sorghum growing countries, India has the largest area followed by Nigeria and Sudan.

In India it is grown over an area of 4.09 m ha with total production of about 4.77 m t. Maharashtra, Karnataka, Rajasthan, Tamil Nadu, Madhya Pradesh and Uttar Pradesh are the important sorghum growing states. Maximum acreage and total production are found in Maharashtra.

Economic importance

Sorghum is an important staple food for millions of people in Africa and Asia. The fodder and stover is fed to millions of livestock providing milk and meat. It is also used as industrial raw material in various industries in U.S.A. and other developed countries. Sorghum grain is eaten by human beings in India either by breaking the grain and cooking it in the same way as rice or by grinding it into flour and preparing 'chapaties'. To some extent it is also eaten as parched and popped grain. This grain is also fed to cattle, poultry and swine. It is also used as feed for dairy cattle, poultry and swine. It can also be used as a source for production of ethanol. Because of Xerophytic features, it is called as "Camel of crops".

Climate

Sorghum thrives well under warm climate. It is very hardy in nature and can withstand high temperature, drought and sometimes short duration excess soil moisture conditions. It is grown from sea level to as high as 1500 m above mean sea level. As such, it is grown from arid regions of Western and Southern India to high rainfall areas of North eastern India. In North and North-east India, it is grown as a *Kharif* crop and in South India, it is grown as *Kharif* as well as *Rabi* crop. The minimum temperature for its germination is 7-10°C, and a temperature range of 26-30°C for its optimum growth.

Soil

Sorghum is grown on a variety of soils in India. Soils with clay loam or loam textured soils with good water retention capacity are the best suited for sorghum. It does well in pH range of 6.0 – 8.5. It may tolerate mild acidity or salinity. The black cotton soils of Central India are very good for its cultivation.

Varieties and hybrids in Tamil nadu

Hybrid	Varieties
CSH 1, 5, 6, 9, 10, 11, 14, CSH19R, CSH 13 R, CSN 15 R, SVP 84, CSH-17 TNAU SH CO 5 TNAU SH CO 4	CSV-13, 14, 15,24SS, 27, SVP 462, SVP 1359, CO(S)-28, K-11, K-12, CO-26, BSR-1, APK-1, Paiyur-1, 2, CO(S)-30, K Tall, Swati, CSV 26, 27, 31 R, 33,

Season

In India sorghum is grown in following three seasons:

1. *Kharif* : June-July
2. *Rabi* : Mid-September-Mid-October
3. Summer : Mid-January-Mid-February

In northern India sorghum is sown only in *Kharif* season. In irrigated areas, first week of July has been found most suitable for sowing of most hybrids and improved varieties. Under un-irrigated conditions, sowing should be done preferably within a week of the onset of first monsoon showers: Timely sown crop escapes the damage due to shoot fly and midge. Late planting may not fit well in multiple rotations. *Rabi* sowing is done mainly in Maharashtra, Karnataka and Andhra Pradesh. *Rabi* sowing should be done from the second fortnight of September to the middle of October. Summer crop of sorghum is sown in the month of January and February in irrigated areas of Tamil Nadu, Andhra Pradesh and some areas of Karnataka.

Field preparation

In a well-prepared field, sorghum seeds should be drilled. The first ploughing should be done with soil turning plough so that 20-25-centimeter-deep soil may become loose. It should be followed by two to three harrowings or three to four intercrossing ploughings with country plough. Thereafter planking should be done to break the clods and to level too field. In black cotton soil area, if the land is badly infested with weeds, ploughing followed by harrowings is usually practiced, but where land is free from weeds or with few weeds, the land is cultivated only with blade harrow.

Seeds and sowing

Seed rate and spacing

To maintain optimum plant population, 15 kg seed per hectare is considered sufficient. There should be 1, 50,000 plants per hectare to attain maximum yield. The seed should be sown in rows 45 cm apart. Plant to plant distance should be 15 cm. Seed should be sown at a depth of 3-4 cm. It should not be sown more than 5 cm deep in any case.

Seed treatment

Seed treatment with Carbendazim at the rate of 2 g per kg seed is recommended. Soaking of sorghum seeds in 2% KH_2PO_4 or 1% $CaCl_2$ for 6 hours and then dried back to the original moisture content in shade is found beneficial under rainfed condition.

Sowing method

Sorghum is sown either by broadcasting or sowing behind the plough. Seeds of new hybrids and varieties should always be sown in lines for obtaining higher yield. Sowing in rows is common in black cotton soil.

Thinning

To maintain desired plant population in sorghum, thinning is very important. Ensure 15 cm plant to plant spacing in a row by thinning out extra plants at two stages. First, thinning should be done 10-15 days after emergence and second, when crop is 20-25 days old. All disease and insect infested plants should be removed while thinning.

Nutrient management

The recommended dose of fertilizer is 100-120 kg nitrogen, 50 kg P_20_5 and 40 kg K_20 per hectare for hybrids and improved varieties of sorghum under irrigated condition. Half dose of nitrogen and total amount of phosphorus and potash should be applied at the time of sowing. The basal dressing can be done with the help of fertilizer-cum-seed drill. The remaining half quantity of nitrogen should be top dressed after 30--35 days after sowing. In light soils top dressing should be done in two splits. Half of the above dose should be applied in case of local varieties for better results.

Under Tamil Nadu condition, blanket recommendation of 90 N, 45 P_2O_5, 45 K_2O kg/ha is followed. Application of half the dose of N and full dose of P_2O_5 and K_2O basally before planting and remaining 50 per cent of N along the furrows on the 15th day of planting is recommended. A quantity of 12.5 kg/ha of micronutrient mixture with enough sand to make a total quantity of 50 kg has to be applied over the furrows and on top one third of the ridges. If micronutrient mixture is not available, 25 kg of zinc sulphate can be mixed with sand to make a total quantity of 50 kg and applied on the furrows and on the top one third of the ridges. Soil application of Azospirillum at 10 packets (2 kg/ha) after mixing with 25 kg of FYM + 25 kg of soil may be carried out before sowing/planting.

In the rainfed areas, application of farm yard manure or compost at the rate of 10 to 15 tonnes per hectare improves the water holding capacity and microbial activities in the soil, besides providing essential nutrients to the crop. Farmyard manure or compost should be added in the field at the time of last ploughing. Under rainfed condition, quantity of fertilizer should be reduced to half of the irrigated and the entire quantity should be applied 10 cm deep in soil at the- time of sowing.

Water management

In India, 90% of sorghum is grown as a rainfed crop. Under prolonged dry spell condition, irrigations should be given during flowering and grain filling stages which are most critical stages for irrigation. Suitable drainage conditions should be provided for the removal of excess rain water from the field. For successful cultivation of sorghum, it requires around 400 mm of water. Irrigation at 75% of depletion of available moisture in 0-30 cm soil layer during *kharif* and 50% during *rabi* and summer is recommended. Sorghum crop irrigated at IW/CPE of 0.4 during *kharif* and 0.8 during *rabi* gave higher yield.

Weed management

During initial period of growth, sorghum crop suffers severe competition from weeds and if the weeds are not brought under control at the right time, there is 20-60 per cent reduction in yield. Application of Atrazine at the rate of 0.5 to 1.0 kg active ingredient per hectare as pre-emergence herbicide is recommended to control weeds at initial stages. The herbicide should be mixed in 500 litres of water and evenly sprayed on the soil surface just after sowing with enough moisture in the soil. Field should not be disturbed for initial 3-4 weeks. To manage the parasitic weed striga in sorghum filed, application of Atrazine in conjunction with 2,4-D is recommended.

Mixed or intercropping of sorghum

Growing of grain sorghum with legumes such as blackgram, greengram, groundnut, soybean and pigeonpea or with oilseed crops like sesamum in various proportions has been a common practice especially in rainfed conditions. Cultivation of fodder sorghum with legumes *e.g.*, cowpea, blackgram, greengram, guar, moth etc. is a customary practice because the legume mixture enriches nutritive value of fodder, increases fodder yield per unit area and also minimizes ill effects of sorghum on soil.

Cropping system

Sorghum is grown in sequence with wheat, pea, bengalgram, potato in North India and Cotton, ragi and tobacco in South India

Ratooning of sorghum

After harvesting of the sorghum crop, the stubbles are left in the field which grow into a crop and give good yield with proper management of the ratoon crop and saves the cost of field preparation, seeds and sowing etc. However, ratoon may harbour more insect pests and diseases as compared to the main crop. The following practices should be adopted for an ideal ratoon crop of sorghum.

Harvesting of the entire first crop should be done at one time when grains are ripe, but stems and leaves are green and not dead ripe. The main crop should be harvested leaving 15 cm stubbles.

The left-over leaves, broken parts of stems and other trash should be removed from the field and burnt to avoid multiplication of insect pests and diseases. Only two healthy sprout stubbles should be allowed to grow and the remaining ones should be removed.

Short duration and the ratooning varieties like CSH-I, CO 25, CO 26, Hybrid CSH 5, K Tall should be selected for the second or ratoon crop.

Ratoon crop needs irrigation as it is a post-monsoon crop. Irrigate the field on 3rd or 4th day after cutting. Subsequently irrigation should be done once in 7 - 10 days and irrigation should be stopped on 70 - 80 days after ratooning. Therefore, it must not suffer due to moisture stress especially from boot leaf to grain filling stages. If the ratoon crop is to be grown on residual soil moisture, all care should be taken to conserve as much moisture *in situ* as possible and subsequently measures be adopted to check the loss of water from the field other than the normal transpiration losses. If possible, harvested water should be recycled at the critical stages of the crop. This practice may succeed in heavy soils with good water holding capacity, but yield will be less than the irrigated crop.

An application of 60-80 kg N/ha in two splits-first soon after the first irrigation before earthing, and the second within 30-35 days of the first application should be done. Ratoon crop matures earlier in 80-85 days after sowing. It should be harvested timely to avoid loss of grains. One needs to be vigilant against attack of insects/other pests including weeds, and disease and their timely prevention and control.

Harvesting and threshing

The high yielding varieties of sorghum mature in about 100-120 days after which they should be harvested without waiting for the dead ripe stage of stems and leaves. In India, generally two methods of harvesting *i.e.,* either stalk cutting or cutting of ear heads by sickles are employed. In case of stalk cut method, the plants are cut from ground level, bundled and stacked on the threshing floor for few days to dry before removing the ear heads. In the latter case the ear heads are piled on the threshing floor for few days to dry before threshing.

Threshing of ear heads is done either by beating them by sticks or by trampling them under bullock feet or under tractor tyre. The threshed grains should be cleaned and dried in sun to a moisture content of 13-15% for safe storage.

Yield

The average yield of sorghum grain is about 5 t/ha of hybrid and improved varieties and about 10-15 t/ha of dry stover under irrigated conditions and about 2.5 t/ha of grain and 8-10 t/ha of dry stover from a rainfed crop.

Sorghum injury

The after effect of sorghum roots and stubbles on succeeding crop of sorghum is called as 'sorghum injury'. This condition lasts for few months or until the sorghum residues decay. Of course, sorghum stubble takes longer time for decomposition than the other cereals being more fibrous. During the process they retain or remove considerable moisture. While decaying lots of microbes build up (sugar content more) and they temporarily lockup N. Sorghum injury can be overcome by additional N application and incorporation with leguminous green manure.

Pearl Millet (Bajra)

(Pennisetum typhoides L.)

Origin

The primary centre of origin of pearl millet is Africa (Southern margins of Saharan Central highlands in Africa) from where it spread to China, India and other countries.

Distribution

Pearl millet is annually grown cereal cultivated in more than 29 m ha in the arid and semi-arid tropical regions of Asia, Africa and Latin America. In Asia and Africa, it is grown as a principal crop replacing sorghum to some extent on sandier soils of dry areas. It is an important cereal crop of India, Pakistan, China, Sudan, Egypt, Arabia Russia and other countries of South-East Asia.

In India, it is one of the important millet crops which flourishes well even under adverse weather conditions. It is the most drought-tolerant crop among the cereals and millets and it provides staple food for the poor in a short period in relatively dry tracts of the country. Pearl millet is grown almost everywhere in the country except the high rainfall areas of Bengal, Assam and North East regions. In India, it is cultivated in an area of 6.93 m ha with a production of 8.61 mt and the productivity is 1243 kg / ha. Among the states, Rajasthan ranks first in area and production, followed by Uttar Pradesh and Maharashtra.

Economic importance

Pearl millet is one of the major coarse grain crops and it is considered to be a poor man's food. It provides staple food for the poor in a short period in the relatively dry tracts of the country. Although pearl millet was developed as a food crop and is still primarily used this way in Africa and India, its grain is most likely to be used for animal feed in the U.S. Pearl millet grains are eaten cooked like rice or 'chapaties' are prepared out of flour like maize or sorghum flour.

Pearl millet is a source of crunchy staple food, cattle and poultry feed, cattle fodder, bedding, thatching and fencing in Arid and semi-arid tropics. It is also used as feed for poultry and green fodder or dry *karbi* for cattle. The grain of Pearl millet is superior in nutritive value to sorghum grain but inferior in feeding value.

Climate

Pearl millet is a quick growing warm weather crop suitable for areas having 40-75 cm of annual rainfall. Usually, Pearl millet is grown in those areas where it is not possible to grow sorghum because of high temperatures and low rainfall. It has a high degree of resistance for drought conditions. It is generally grown during South West monsoon or *kharif* season, but with the availability of photo-insensitive varieties, it can be grown in *kharif, rabi* and *zaid* if irrigation facilities are available. During the vegetative phase, the crop needs wet weather, light showers, and bright sunshine. The crop may tolerate drought, but cannot withstand a high rainfall of about 90 cm or more and water logging. It cannot tolerate frost also. The best temperature for the growth of Pearl millet is between 20 and 28°C.

Soil

Pearl millet can be grown on a wide variety of soils, but being sensitive to water-logging, it does better on well-drained sandy loams. It is grown successfully on black cotton soils, alluvial soils and red soils of India. Light textured soils having low fertility, good drainage and neutral to mild salinity, are the best for the crop. It is sensitive to acidic soils. The soil should be deep and free from stones and concretions. It is being grown on a wide variety of soils, namely, black cotton soil alluvial soil and red soils of Deccan plateau.

Varieties and Hybrids

Some other Hybrid varieties identified for different zones which are medium early, higher yielders and resistant to Downy mildew are: MH-518, MH-597, MH-515, MH-552, MP-258 and IP-266. HHB 299 (Iron & Zinc rich hybrid), AHB 1200 (Iron rich hybrid).

The hybrids suitable for Tamil Nadu are COH (Cu)-8, Nandi-35, PAC-903, X-7, GK-1004, and B-2301.

The variety cultivated in Tamil Nadu is Co (Cu)-9, CO7, ICMV 221, Co (Cu)-10, HHB 299 (MH 2076), AHB 1200 Fe (MH 2072, NBH 4903 (Balwan), PBH 306 (MH 1962), Nandi 75, 86M01 (MH 1790), 86M01 (MH 1790), Pratap (MH 1642), 86M 53 & 64, PB 727 (Proagro9555), GHB 526 & 558, Pusa 23, 322,

444, & 763, RHB 58, ICMH 356, EKNATH 301, MLBH 104, HHB 67, VBH 4, MH 179 & 182.

Season

Pearl millet is chiefly grown under rainfed conditions during monsoon (*Kharif*) season in North India. However, it can be grown round the year in areas of assured water supply in South India.

Field preparation

Pearl millet requires well pulverised seedbed with fine tilth which is free from clods as the seeds are of very small size. Summer ploughing with mould-board plough and solarization of the soil during summer is a must for seedbed preparation for pearl millet. With the onset of monsoon, the field should be harrowed, ploughed by local plough, or with tractor drawn cultivator followed by planking to pulverize the soil and to destroy the weeds. Farmyard manure should be applied 10 days before starting to prepare the land. Adequate moisture in the seedbed at the time of sowing is conducive to good germination. Formation of ridges and furrows in 45 cm apart is followed.

Seeds and sowing

Seed rate and spacing

Pearl millet requires 5 kg seeds per hectare to ensure optimum plant population. By adopting the spacing of 45 cm between rows and 10 cm between the plants, optimum population of 200,000-2,22,000 plants per hectare. could be achieved.

Method of sowing

Pearl millet is generally sown behind plough or by broadcast method. However, this method of sowing leads to poor germination and consequently poor yield. Sowing pearl millet with seed drill is the best method. It not only ensures better germination but uniform plant population as well. Optimum depth of sowing is around 2 to 3 cm.

Sowing time

Sowing time in most of the states is June-July

In northern India, pearl millet is grown mainly in *Kharif* season. As a *Kharif* crop, optimum time of sowing is first fortnight of July. However, it can be raised round the year in areas of assured water supply in south. In case, sowing is delayed there is a drastic reduction in yield due to more incidence of diseases like downy mildew and ergot, restricted vegetative growth of the crop, high rate of mortality and poor grain setting.

Transplanting of seedlings

In case of delayed sowing due to several unavoidable reasons such as late onset of monsoon, heavy and continuous rain during the optimum sowing time or late harvesting of preceding crop, transplanting can be recommended as compared to direct sowing which produces the following distinct advantages:

Transplanted crop matures earlier and low temperature late in the season has no adverse effect on grain setting. It also produces more tillers and ears owing to better growth. Optimum plant population is ensured. Transplanted crop gets a better start because the three weeks old seedlings are able to withstand frequent rains. Downy mildew infected seedlings are rejected at the time of transplanting itself.

Nursery

Nursery area of 500-600 square meter is sufficient in nursery to get seedlings for one hectare. Raised bed of size 1.2 m x 7.5 cm is prepared in 10 cm apart and 15 cm depth. Application of 750 kg of FYM or compost is done and incorporated by ploughing. About 25-30 kg calcium ammonium nitrate is applied uniformly over the nursery. To protect the seedlings from shoot fly infestation, phorate 10 G 180 g or Carbofuran 3 G 600 g mixed with 2 kg of moist sand and applied on the beds. About 2 kg seed of pearl millet is sown in the beds. The seedlings are uprooted and transplanted after three weeks. While uprooting the seedlings, the nursery should be kept wet to avoid root injury. During transplanting the field should be irrigated to help the seedlings to establish themselves. One seedling per hole is transplanted in rows keeping 50 cm space between rows and 10 cm space between plants.

Nutrient management

General recommendations are 100-120 kg nitrogen, 40-60 kg P_2O_5 and 30-40 kg K_2O per hectare. Half of the above doses should be applied under rainfed conditions. Half dose of nit-rogen and full doses of phosphorus and potassium should be applied at the time of sowing in furrows approximately 3-5 cm below the seed through seed-cum-fertilizer drills. The remaining nitrogen is top-dressed in two splits, one at the time of thinning (three to four weeks after sowing) and rest at ear formation stage. Under rainfed conditions foliar spray of 3 % urea is also recom-mended.

In Tamil Nadu, 12.5 t/ha of FYM is applied. Soil application of Azospirillum @ 10 packets per ha (2000 g) is recommended. The blanket recommendation of 70:35:35 kg N, P_2O5, K_2O/ha is followed for all varieties. For hybrids, 80 kg N, 40 kg P_2O5 and 40 kg K_2O per ha are applied. Application of 50 per cent of the recommended nitrogen and full dose of phosphorus and potassium as basal and the remaining 50% of nitrogen 15 days after transplanting for transplanted crop and 30 days after sowing for direct sown crop is recommended. Micronutrient mixture at 12.5 kg/ha mixed sand to make 50 kg is applied on the surface just before planting/after sowing and the seeds are covered.

Water management

Pearl millet is predominantly grown as rainfed crop (91%), irrigation is hardly needed. However, by adopting proper soil-water conservation measures including efficient drainage facilities successful cultivation of pearl millet under rainfed conditions is possible. Early vegetative, heading and flowering stages are water sensitive critical stages of pearl millet. Pearl millet is badly affected by water logging. Therefore, rain water should not be allowed to stay in the field for more than a few hours.

Weed management

The first 30-45 days after sowing of pearl millet are most critical for control of weeds. Timely weeding 30-35 days after sowing by hand or by wheel hoe or hand hoe is a must. If hand weeded, stirring of the top soil between the rows is necessary. Pre-emergence application of Atrazine or Propazine @ 0.5 kg/ha in 500 litres of water controls most of the monocot and dicot weeds. Then, one hand weeding on 30-35 days after transplanting may be given. If herbicide is not used, hand weed on 15th day and again between 30 and 35 days after transplanting may be done.

Cropping system

In North Indian conditions some of the crop rotations are as follows:

- Pearl millet - wheat - green grarn / urdbean (Irrigated)
- Pearl millet - wheat / pea / potato / mustard / vegetables (Irrigated)
- Pearl millet - toria – wheat / potato (Irrigated)
- Pearl millet - barley/chickpea / lenti / linseed / toria / mustard - rainfed on conserved moisture.

In rainfed areas, intercropping of pearl millet with groundnut, soybean, sesamum, urdbean, and mungbean, pigeonpea or castor are more remunerative.

Harvesting and threshing

Harvesting is done by cutting the entire plant from ground level and removing the ear head after few days drying or cutting the earheads only and drying them well for few days before threshing. Grains are separated either by beating the ear heads with sticks or by trampling by bullocks or under tractor wheels. The threshed grains are cleaned and dried in sun to bring the moisture content to 14% for safe storage.

Yield

The average yield of pearl millet is about 3-4 t/ha of hybrid and improved varieties under irrigated conditions and about 10 t/ha of dry stover and about 1.5-2.0 t/ha grain and 7.0-7.5 t/ha dry stover yields from a rainfed crop.

Questions

I. Choose the best from the choices given

1. The crop which is called camel of crops is

   a) Pearl millet b) Sorghum

   c) Finger millet d) Varaghu

2. To avoid sorghum poisoning in animals, feeding of sorghum shoud be avoided

   a) Upto 35 days b) Upto 40 days

   c) Upto 45 days d) Upto 50 days

3. Sorghum injury can be overcome by

   a) Repeated ploughing b) Applying P_2O_5

   c) Applying N d) Applying K_2O

4. The highest area of Pearlmillet is in

   a) Tamil Nadu b) Rajasthan

   c) Uttar Pradesh d) Andra pradesh

5. The important millet crop which flouishes well even under adverse weather conditions are

   a) Sorghum b) Pearl millet

   c Finger millet d) Kodo millet

II. Fill in the blanks

1. Seed rate for pearl millet is __________
2. __________ is called as camel of crops.
3. Pearl millet is native of _________
4. In sorghum row spacing of _____ cm is advised.
5. Water requirement of sorghum is about _______ mm.

III. Write short notes

1. Sorghum injury
2. Sorghum poisoning
3. Weed management in pearl millet
4. Method of sowing in pearl millet
5. Importance of pearl millet

8

Finger Millet and Minor Millets - Origin, Geographic Distribution, Economic Importance, Soil and Climatic Requirement, Varieties, Cultural Practices and Yield

FINGER MILLET *(RAGI) (Eleusine coracana)*

Origin

Finger millet (Ragi/ Mandua) is the principal small millet grown in South Asia. It is believed to be native of India (De condole, 1886) but according to Vavilov (1926) it originated in Abyssinia.

Distribution

Finger millet is widely cultivated in India, Africa, Ceylon, Malaysia, China and Japan. It is an important minor millet grown in India. In India, it is cultivated over an area of 1.19 million hectares with total production of about 1.98 million tonnes and the average productivity of 1662 kg/ha. It is a staple food crop in many hilly regions of the country. This crop is cultivated up to an altitude of 2100 metres above the sea level. It is grown mostly in Karnataka, Uttarkand, Maharastra, Tamil Nadu, Odisha and other parts of India.

Economic importance

It is known as nutri grain due to its low fat, high protein, minerals and vitamins. It serves as the staple food for the people in the arid and semi-arid tropics. It is grown both for grain and forage. In northern hills, grains are eaten mostly in the form of 'chapaties'. In south India grains are used in many preparations like cakes, puddings, sweets, etc. Germinating grains are malted and fed to infants also. It is also good for pregnant women. It is used for malting and brewing. Malted finger millet is used in infant food and in milk thickness formulations. Finger millet malt brewed with an equal proportion produces beer of commercial standards. Its grain has nutritive value and is considered fit for the people who do hard physical work. It is a nutritive food for adults of different ages. It is good for persons suffering from diabetes. The green straw is suitable for making silage, which is sweet smelling and consumed by cattle without any wastage. It can be grown both as the dry land crop and also under irrigation. The crop is free from pest and diseases. Its straw is used as fodder.

Climate

Finger millet is grown in tropical as well as sub tropical climate upto an altitude of 2100 m. Preferred altitude for this crop is 1000-1800 m. It is a heat loving plant and for germination, the minimum temperature required is 8-10°C. A mean temperature range of 26-29°C during the growth is the best for proper development and good crop yield. Crop yield reduces at temperature below 20°C. It is one of the hardiest crops suited for dry farming. It can be grown under condition of low rainfall and can withstand in drought, reviving again with a good shower of rain with remarkable vigour. It is cultivated as dryland crop where rainfall ranges 500 to 750 mm. The yield of finger millet is directly related to amount of rainfall received from June to October. It has the ability to tolerate salinity among cereals.

Soil

Finger millet can be grown well in red soil, alluvial soil and black cotton soil. It can be grown in light soils, loamy soils, sandy loam soils with good drainage. Heavy soils with poor drainage are not mostly preferred. It can tolerate salinity and higher pH.

Varieties

A number of short and medium duration varieties of finger millet are available for cultivation in *kharif*, *rabi* and *summer* seasons. Growing short duration varieties in rainfed upland will permit double cropping of finger millet with pulses or sesame. The varieties suitable for important states are given below:

State	Varieties
Andhra Pradesh	AKP-2, AKP-3 and AKP-7 VR 936, PPR 2700 (Vakula), Srichaitanya (VR 847)
Karnataka	Poona, Kaveria, Hansa, Annapurana, Hagari 1 and Godavari (PR-202), KMR 340, OEB 532, KMR 204, GPU 66,Indaf 5, Indaf 9, L 5, GPU 48, GPU-45, MR 6, GPU-28, GPU-26, MR-1, MR-2, HR-911, BM9-1, Cauveri, Hamsas, Annapurna, Hagari 1, Hagari 2, Shakti Hullubele, Hasta (Indaf 7) and PR-202 (Godavari)

Tamil Nadu	CO RA (14), CO 1, CO 2, CO 3, CO 4, CO 5, CO 6, CO 7, CO 8, CO 9, Co 10, CO 11, CO 12, CO 13, K 1, K 2, K 5, K 6, K 7, TRY 1, Paiyur 1, Paiyur (RA) 2, GPU 28, GPU 48, Indaf 5, Indaf 7, Indaf 9, PR-202
Kerala	PR-202, K-2, CO-2, CO-7, CO-8, CO-9, CO-10
Maharastra	B 1, E 31 and A16, KOPN 235
All over India	Sarada, EC 4840, IE 28, VL 376, VL 352

Seasons

Finger millet can be grown throughout the year as *kharif* crops, as a *rabi* crop with light irrigation and as a *summer* crop with moderate irrigation. Flowering will be delayed if it coincides with cold weather. Amongst the *rabi* millets, finger millet is the most important crop. Because of its low irrigation needs, it has great scope for cultivation in the *rabi* and *summer* seasons in the tail ends of canals and in the command areas of minor irrigation projects in multiple cropping sequences. Finger millet, though predominantly a *kharif* crop, can be cultivated throughout the year if irrigation is available. *Kharif* season main crop is sown during May or June. The irrigated crop in *rabi* season in Karnataka, Tamil Nadu and Andhra Pradesh is sown in September and October. In higher hills of North India, the optimum time of sowing for finger millet is first fortnight of June.

Field preparation

The land is prepared thoroughly and ploughing starts from the first shower of rain and repeated with succeeding rain. The first ploughing with mould board plough should be done immediately after the harvest of the previous crop. In south India, where the soil becomes too hard for ploughing, a shallow stirring should be given by bladed harrow or a disc harrow. With the onset of monsoon, field should be harrowed or ploughed with local plough two to three times and finally levelled. In south India under rainfed conditions, with every succeeding rain the ploughing is repeated or the 'kunte' is used. Seed bed for finger millet is usually very thoroughly prepared, which should be free of weeds, friable and smooth for better germination and crop growth.

Seeds and sowing

Seed treatment

The seeds are treated with either Thiram 4 g/kg or Captan 4 g/kg or Carbendazim 2 g/kg of seeds. The seeds are to be treated at least 24 hours prior to sowing. Seed treatment with Azospirillum may be done at 3 packets/ha (600 g/ha).

Direct seeded crop

For direct seeding of finger millet, a well pulverized seed bed should be prepared. Five tonnes of FYM or compost / ha should be incorporated well into the soil along with fertilizers before sowing. Sowing the direct seeded crop in lines 20 cm apart is better than the broadcasted one. Seed rate of about 10 kg/ha will be adequate for the line sown crop. The seeds selected for sowing should be free from mixture and under developed seeds. The under developed seeds can be separated out. For the selection of healthy seeds, salt water treatment can be practiced. Generally, broadcasting and drilling methods are used for sowing of direct seeded crop of finger millet.

Transplanted crop

Nursery

For raising seedlings to plant one ha of main field, 12.5 cents (500 m^2) of nursery area near a water source, where water does not stagnate is required. Application of 37.5 kg of super phosphate along with 500 kg of FYM or compost evenly on the nursery area is important. Ploughing two or three times with a mould board plough or five times with a country plough is essential. Nursery area can be divided into small units of 6 plots each of size 3 m x 1.5 m with a spacing of 30 cm between plots for irrigation. Before sowing, shallow rills not deeper than one cm on the beds are formed by passing the fingers vertically over them. A quantity of 5 kg of treated seeds are broadcasted evenly on the beds and the seeds are covered by levelling by hand lightly over the soil. Sprinkle 500 kg of powdered FYM over the beds evenly to cover the seeds which are exposed and compact the surface lightly. Sowing of seeds should be done evenly in the nursery beds early in the month of June. The nursery beds may be irrigated, if necessary. The seedlings will be ready within 18 to 20 days.

Transplanting

After land preparation, manures and fertilizers should be applied. Transplanting should be close. Early varieties should be transplanted 15 cm x 10 cm apart and medium and late varieties 20 x 10 cm a part. It is however, economical to transplant seedlings in plough furrows in which case both early and late varieties may be planted in plough furrows drawn 20 cm apart. Two seedlings may be placed 10 cm apart in the furrows at each hill. The base of the seedlings will be covered by soil when the next furrow is drawn. Shallow planting within 3 cm depth encourages quicker and better tillering.

Nutrient management

Application of 12.5 t/ha of FYM or compost evenly on the unploughed field and then ploughing to incorporate in the soil is recommended. It is better to apply fertilizer as per soil test recommendation. If soil test is not done, the fertilizer may be applied at the rate of 40 kg Nitrogen, 20 kg P_2O_5 and 20 kg K_20 for short duration varieties of 90-95 days and 60 kg N, 30 kg P_2O_5 and 30 kg K_2O per hectare for medium and long duration varieties of finger millet. For rainfed crop, only 20 kg of N and 12 kg each of P_2O_5 and K_2O per hectare may be applied. Nitrogen may be applied in two splits, *i.e.,* 50% as basal and remaining 50% as top dressing at the first hoeing and weeding so as to incorporate the fertilizer into the soil. 12.5 kg of micronutrient mixture has to be mixed with enough sand to make a total quality of 50 kg/ha and applied uniformly.

Weed management

Early weeding of the direct seeded crop is essential for smothering weeds and getting good yields. Herbicides like Butachlor 1.25 kg/ha or Fluchloralin 2 l/ha or Pendimethalin 2.5 l/ha can be sprayed as pre-emergence. If pre-emergence herbicide is not applied, hand weeding twice on 10th and 20th day after transplanting may be followed. For direct seeded crop, post-emergence herbicide 2,4 DEE or 2,4 D Na salt at 0.5 kg on 10 days after crop germination could be applied to control broad leaved weeds.

Water management

Finger millet cannot withstand water logging but it can tolerate better than maize and jowar. The field should therefore, be drained out during the monsoon if necessary. Under water scarce condition, two irrigations at tillering and flowering stages can increase the yield. Application of irrigation water at 50% depletion of available soil moisture is sufficient. Number of irrigations is based on growth phases of crop *viz.,* establishment - 2 irrigations, vegetative up to 25 days - 2 irrigations, flowering - 25-55 days - 3 irrigations, 56 days to maturity - one or two irrigations. Irrigation has to be withdrawn after dough stage.

Cropping system

Finger millet in rainfed conditions is cultivated generally as a mixed crop with sorghum, pearl millet and a variety of oilseeds and pulses. In hilly areas it is grown mixed with soybean. Under irrigated condition, it is grown in rotation with crops like tobacco, vegetables, turmeric, gram, linseed, mustard, etc.,

some of the most prevalent cropping sequences are: finger millet - gram, finger millet - mustard, finger millet - tobacco, finger millet - groundnut, finger millet - sugarcane, finger millet - potato- maize, finger millet - rice.

Harvesting and threshing

Finger millet matures in about 120-135 days depending on the tract and the variety. Harvesting is generally done in two methods. The ear heads are harvested with ordinary sickles and straw is cut close to the ground. Ear heads are heaped for three to four days to cure and then threshed. At some places under rainfed conditions, the whole plant with ear head is cut, heaped and then threshed. Threshing is done by three methods: Treading out the grains under the feet of the bullocks or beating out the grains with the sticks, and moving a stone roller over the ear heads.

The first method is the most common. After thrashing, there is always a small proportion of ear head left unthreshed which has to be beaten by sticks.

Yield

The grain yield of finger millet under irrigated condition is 3 t / ha, but the average yield is only 2 t / ha. Under rainfed condition the grain yield is 1.5 t / ha, but the average yield only 1.0 t / ha

Minor millets

Minor millets (small millets) have received less attention than major millets in terms of cultivation and utilization. Minor millets account for less than one percent of the food grains produced in the world today. Thus, they are not important in terms of world food production, but they are essential as food crops in their respective agro-ecosystems. They are mostly grown in marginal areas or under agricultural conditions where major cereals fail to give sustainable yields.

Minor millets	Botanical Name
Foxtail millet	*Setaria italica*
Kodo millet	*Paspalum scrobiculatum*
Common or prosomillet	*Panicum miliaceum*
Little millet	*Panicum sumatrense*
Barnyard or sawa millet	*Echinochloa crus-galli Echinochloa corona*

Special features of minor millets

Minor millets	Special features
Foxtail millet	Susceptible to both drought and water logging
Kodo millet	Highly drought resistant
Little millet	Highly drought resistant and tolerant to water logging
Proso / Common millet	Highly drought resistant
Barn yard millet	Highly drought resistant and tolerant to water logging

Special characters of minor millets

They grow during adverse soil and climate where the production of other crops is not possible. They are mostly shorter in duration (100 days). Some of them are suitable for contingency plan. Proso and little millet mature in shorter duration and thus they provide food during lean months for tribal people. They have relatively higher nutritional value compared to major cereals. Cultivation is easy and cost of cultivation is less.

Area and distribution

Small millets are distributed in most of the Asia and African countries and part of Europe. Finger millet is the principal crop small millet grown in south Asia. Foxtail millet and prosomillet are important in China and prosomillet is extensively grown in southwestern Russia. In India, small millets are predominantly gown in Madhya Pradesh, Chatishgarh, Maharastra Karnataka, Odhisa, Uttakhand and Tamil Nadu.

In India it is grown over an area of 6.19 lakh hectares with total production of about 4.41 lakh tonnes. Madhya Pradesh, Chhattisgarh, Maharashtra, Uttaranchal and Tamil Nadu are the important minor millets growing states. Maximum acreage and total production are found in Madhya Pradesh. Utharkhand gives the highest average yield per hectare (1.29 t / ha) followed by Tamil Nadu (1.24 t / ha).

Foxtail millet - *Setaria italica L.,*

Foxtail millet is also known as Italian, German Hungarian or Siberian millet. It is generally considered to have been domesticated in eastern Asia, where it has been cultivated since ancient times. The main production area is China, but *S. italica* is the most important millet in Japan and is widely cultivated in India. It is believed to have been one of the five sacred plants of ancient China (from 2700 BC). Because of its short duration it is a suitable crop for growing by nomads, and it was probably brought to Europe in this way during the Stone Age, as seeds abound in the Lake Dwellings in Europe. The height of the plants varies from 1 to 1.5 m and the colour of the grain varies from pale yellow, through orange, red and brown to black. The 1000-seed weight is about 2 g.

Common millet - *Panicum miliaceum L.,*

Common millet is also known as proso millet, hog millet, broom-corn millet, Russian millet and brown corn. This millet is of ancient cultivation. It is the milium of the Romans and the true millet of history. It was cultivated by the early Lake Dwellers in Europe. It is believed to have been domesticated in central and eastern Asia and because of its ability to mature quickly was often grown by nomads. The shallow-rooted plant varies in height between 30 and 100 cm. The grain contains a comparatively high percentage of indigestible fibre because the seeds are enclosed in the hulls and are difficult to remove by conventional milling processes. The 1000-seed weight is about 5 g (varying between 4.7 and 7.2 g). Common millet is particularly suited to dry continental conditions and grows in more temperate climates than other millets.

Little millet- *Panicum sumatrense*

Little millet is grown throughout India to a limited extent up to altitudes of 2100 m but is of little importance elsewhere. It has received comparatively little attention. The plant varies in height between 30 and 90 cm and its oblong panicle varies in length between 14 and 40 cm. The seeds of little millet are smaller than those of common millet.

Barnyard millet- *Echinochloa crusgalli (L.) P.B* and *Echinochloa colona (L.)*

Barnyard, Japanese barnyard or sawa millet is the fastest growing of all millets and produces a crop in six weeks. It is grown in India, Japan and China as a substitute for rice when the paddy fails. It is grown as a forage crop in the United States and can produce as many as eight harvests per year. The plant has attracted some attention as a fodder in the United States and Japan. The height of the plant varies between 50 and 100 cm.

Kodo millet - *Paspalum scrobiculatum L.,*

Kodo millet is a minor grain crop in India but is of great importance in the Deccan Plateau. Its cultivation in India is generally confined to Gujarat, Karnataka and parts of Tamil Nadu. Kodo is an annual tufted grass that grows to 90 cm high. Some forms have been reported to be poisonous to humans and animals, possibly because of a fungus infecting the grain. The grain is enclosed in hard, corneous, persistent husks that are difficult to remove. The grain may vary in colour from light red to dark grey.

Small millets- Agro-techniques

Climate

Most small millets except *Setaria italica* are grown in warm regions 35-40°C. It grows well in moderate weather of 26-29°C. Fox tail millet grows well in temperate and as well as tropics. Minor millets require well distributed rainfall of 150-350 mm.

Soil

Minor millets require well drained soils of medium texture with pH of 5 to 8. They are essentially dryland crop on marginal and sub marginal lands in tropical countries. Prosomillet is cultivated in light shallow red soils and relatively sandy soils.

Varieties

Minor millets	Varieties
Foxtail millet- *Kangni*	Arjun, CO 1, CO 2, CO 3, RS 118, CO 4, CO 5, Chitra, nallamala, Nischal, K3, K2, CO 6, Meera, Pratap, PRK 1, Sri lashmi, Gavari (SR 11), SIA, 2622, H2, PS 4 and SIC 3, CO (Te) 7, SiA 3156, SiA 3088, SiA 3085, HMT 100-1, SR 51.
Kodo millet – *Kodra*	CO3, APK -1, PSC 10, PLR 1, Niwas, JNK 101, JNK 364, Jawahar, Kodo 1, Kodo 2, K1, Kodo 41, Kodo 62, PSC 1 and Gukarat Kodara 1, GPUK 3, Kodo 48, kodo 76 and KK2, CO3, TNAU 86, RK 390-25, Indra kodo 1, DPS 9-1, JK98JK, 36JK65, JK 13, 48, 106, 155 & 439, KMV 20
Little millet – *Kutki*	CO3, CO2, CO1, Paiyur 2, Dindori 1, Gujarat Vari, 1, PRC 3, Jawahar Kutki, Jawahar, Paiyur 1, Birsa Gundi 1, Kolab, Tarini (OLM 203) and K1 Paiyur 2, CO4, ATL 1, OLM 217, OLM 208, JK 36
Proso/Common millet – *Cheena*	CO 1, CO 2, CO 3, CO 4, K1, K2, Ram Cheena, Shyam Cheena, Varada, Nagarjuna, sagar, GPAU 8 and Bhawana CO(PV) 5, TNAU 202, TNAU 164, TNAU 151, PRC 1, TNAU 145, CO 5, PR 18, GPUP 21, GPUP 8
Barn yard millet- *Sawan*	CO 1, K1, K2, MDU 1, VL Madira 8, Gujarat Banti 1, VL Madira 21, VL Madira 172, VL Madira 181, RAU 3, Anurag, Chandana, CO 2, VL Madira 207, ER 64, PRJ 1, RAU 11

Season

Minor millets are being cultivated during rainy season, mostly as rainfed crop. Usual sowing time for *kharif* crop is June and for *rabi* crop is October. Prosomillet is extensively grown unirrigated as a late monsoon crop on poor soils.

Field preparation

Since the seed is small, minor millets require thorough land preparation for adequate stand establishment. Light soils are given 2 to 3 ploughings followed by a couple of harrowings with blade harrows. In case of black soils, repeated blade harrowings will be done until a fine seedbed is obtained.

Seeds and Sowing

Seed rate and spacing

Minor millets	Seed rate (kg/ha)	Spacing (cm)
Foxtail millet	Drilling – 5 kg Irrigated transplanting – 3 kg	22.5 to 30 cm between rows
Kodo millet	Drilling and broadcasting- 10-12 kg	22.5 to 30 cm between rows
Little millet	Drilling and broadcasting – 10 kg	22.5 to 30 cm between rows
Proso / Common millet	Drilling and broadcasting – 10 kg	22.5 to 30 cm between rows
Barn yard millet	Drilling and broadcasting – 10 kg	30 cm between rows

Seed treatment

Seeds are treated with bavistin and coated with Azotobater or Azospirillum that protects crop from seed borne diseases, produces vigorous plants and results in better yields. On an average about 8 - 10 kg/ha seed is required for broadcasting and 6-8 kg/ha for line sowing. Growing improved varieties is always better for obtaining higher productivity.

Sowing time and method

Since the crop is grown as rainfed on marginal lands, sowing should be done immediately after the onset of monsoon (in first week of July). Delayed sowing causes drastic reduction in yield. Farmer's practice has been to broadcast the seed, mix it in the soil followed by planking- but line sowing at 22.5 to 30 cm row spacing always gives higher grain yield. Seeds, being very small, must be mixed in sand for a uniform distribution in the field.

Minor millets	Sowing time	Sowing method
Foxtail millet	*Kharif* – June Late *Kharif* – August Summer irrigated –Jan-Feb	Broadcasting and drilling Transplanting
Kodo millet	June -July	Broadcasting and drilling
Little millet	June -July	Broadcasting and drilling
Proso millet	September – October	Broadcasting and drilling
Barn yard millet	*Kharif* – June *Rabi* - July	Broadcasting and drilling

Thinning: Thinning should be done during 1st hand weeding.

Nutrient management

Under dryland conditions, minor millets may not receive manures and fertilizers. Application of organic manures at 5 to 10 t/ha at the time of last ploughing is recommended. However, minor millets respond to inorganic fertilizer at the rate of 20-40 kg N, 10-20 kg P_2O_5, 10-20 kg K_2O. Under irrigated condition, it responds to higher N dose up to 60 kg/ha. Inorganic fertilizers are mostly applied as basal dose at the time of sowing.

Irrigation management

Minor millets are mostly grown during rainy season as rainfed crop. If rains are well distributed the crop does not need irrigation. However, one or two irrigations during critical periods especially flowering and grain development stages are found to be beneficial. Mulching may be done for conserving moisture and higher productivity.

Weed management

Minor millets require a weed free period of 20-35 days after sowing for better crop growth. Two intercultural operations during 20 and 35 days after sowing is important to control the weeds. In case of non availability of labour, application of Isoproturan @ 0.5 kg or Pendimethalin @ 0.5 kg/ha on 3rd day after sowing is recommended. To control the broad-leaved weeds, application of post emergence herbicide 2,4 D Na salt at 0.5 kg on 20-25 DAS can also be followed.

Cropping system

Under dryland condition, it is mixed with cotton, maize, sorghum, redgram and other pulses, since they are *Kharif* season crops. In r*abi* they are mixed with rape seed, mustard, gram, lentil, linseed, barley etc. under rainfed condition.

Harvest

Harvesting is done by removal of ear heads from the stocks after ear heads are fully matured. Harvested ear heads are sun dried for a day or two and threshed by trampling under the feet of farm animals using stone roller or beating with wooden sticks. Winnowing is done to remove the chaff grains and dried to 14 % moisture.

Yield

Minor millets	Average yield (kg/ha)	
	Grain	Straw
Foxtail millet	Rainfed - 800-1200 Irrigated -1500-2000	Rainfed -1500 Irrigated -2000
Kodo millet	Rainfed - 600-800	1500
Little millet	Rainfed - 400	-
Proso /Common millet	Rainfed - 400-500	1000-1500
Barn yard millet	Rainfed - 400	1000

Questions

I. Choose the best from the choices given

1. The minor millet which has long duration is
   a) Samai b) Panivaragu
   c) Varagu d) Kudiraivali
2. The minor millet which withstands both flood and drought
   a) Kudiraivalli b) Samai
   c) Panivaragu d) A & B
3. The millet recommended for diabetic people is
   a) Sorghum b) Ragi
   c) Pearlmillet d) Kodo millet
4. Common name of finger millet is
   a) Sorghum b) Pearl millet
   c) Ragi d) Sawan
5. Tillering capacity is high in
   a) Kodo millet b) Kuthiraivali
   c) Both a & b d) None

II. Fill in the blanks

1. Among the small millets, _________ millet is the longest duration dry land crop.
2. Seed rate of finger millet is ____
3. Small millets are also called as ________

4. ________ millet can be cultivated in saline soils.
5. The grain yield of little millet under rainfed is_______

III. Write short notes

1. Importance of small millets
2. Special characters of minor millets
3. Cropping system involving finger millet
4. Uses of finger millet
5. Minor millet based cropping systems

9

Pigeonpea - Origin, Geographic Distribution, Economic Importance, Soil and Climate Requirement and Yield

PULSES

Pulses are seeds of leguminous plants used as food. They produce dal rich in protein. The important pulse crops are greengram, blackgram, redgram, bengal gram, soybean, cowpea, peas, lentil and lathyrus. The total area under pulses in India is 29.81 m ha with a production of 25.42 m t and the productivity is 853 kg / ha. Pulses can grow both in rainfed and irrigated conditions.

The productivity of pulses in India is generally low and the causes for low yield of pulses are: Low harvest index as pulses are C3 plants, indeterminate growth habit, non-synchronous flowering and fruiting, low sink potential, non-responsiveness to fertilizers, high weed infestation, improper sowing time and non-availability of good quality seeds at the right time.

Pigeonpea (*Cajanus cajan* L.)

It is commonly known as Arhar. It is an important and old crop of this country. It is the second most important pulse crop only after chickpea. Pigeonpea is a crop primarily of India, though there are substantial areas in Africa especially in eastern Africa where pigeonpea is grown. In India it is cultivated in an area of 4.4 m ha with a production of 4.23 m t and the productivity is 967 kg / ha. It is also becoming increasingly important in central and South America. Though pigeonpea is grown in a wide range of agro-ecological situations, its deep rooting and drought tolerant characters make it especially useful crop in the area of low and uncertain rainfall and on the lighter soils. Red gram is sometimes grown as a sole crop, but more typically, it is grown in relatively complex systems where it is intercropped, or mixed with other crops. Traditionally long duration varieties of pigeonpea which take about 240-270 days to mature are of low yield potential and susceptible to different diseases and pests. These varieties are also damaged by frosts which frequently occur in most parts of northern India in the month of December and January. Recently some short duration varieties of about 130-160 days have been developed which have high yield potential (2-3 t / ha) and harvested by end of November.

Origin

Pigeonpea is the most widely grown crop in the country and has been under cultivation for over three thousand years. It has been reported to occur in wild state in the upper region of Nile River and the coastal districts of Angola in Africa. Therefore, Africa may be its original place and from where it might have been introduced to India. However, according to Vavilov, India is the place of origin of pigeonpea.

Distribution

Pigeonpea is grown throughout the tropical countries of the world especially in more arid regions of Africa, West Indies, Srilanka, Australia and Malaya. The prominent pigeonpea growing states in India are Maharashtra, Karnataka, Madhya Pradesh, Telangana, Uttar Pradesh, Gujarat, Jharkhand, Andhra Pradesh and Odisha.

Area and production

Pigeonpea is grown on an area of about 4.44 million hectares with the production of 4.23 million tonnes in the country. The national productivity of pigeonpea is 967 kg per hectare. It is the second most important crop after chickpea among different pulse crops in the country. Maharashtra has the largest area and production of pigeonpea.

Economic importance

Pigeonpea is mainly used as 'dal'. It is rich in iron, iodine and the essential amino-acids like lysine, cystine and arginine. It is also used as ration for milch cattle. Its straw is also palatable and green leaves may be used as fodder. Sticks of pigeonpea are used for various purposes such as thatch, basket making, etc. Pigeonpea being a leguminous plant is capable of fixing atmospheric nitrogen and thereby restore lot of nitrogen to the soil. Its deep root system helps in extracting the nutrients and moisture from deeper soil layers. Deeper root system also helps in breaking the plough pans and help in improving soil structure. Hence this crop is called as biological plough. Extensive ground cover of pigenpea prevents soil from wind and water erosion, encourages infiltration, minimizes sedimentation and smothers weeds. It is also grown on mountain slopes to utilize the poor soils and protect it from erosion.

Varieties

At present emphasis is being laid on the development of short duration high yielding varieties permitting second crop of wheat or other *rabi* crop in the winter. These varieties mature in about 130-180 days. They are harvested by December and escape the risk of frost in the end of December and January which commonly occur in northern India. Earlier longer duration varieties taking 240-300 days to maturity were grown, occupying field for whole of the year. These varieties were usually damaged by frost in the month of December and January under north Indian conditions.

Some of the important varieties cultivated in north india are Prabhat, UPAS 120, Pusa Ageti of early duration (120-130 days), Paras, Mukta, Sharda, Pusa 74, Pusa 84, Laxmi of medium duration (140-170 days) and NP (WR) 15, T7, T 17, Pusa 33 and SA 1 of late maturing (more than 180 days).

Soil

Pigeonpea requires light textured, well-drained soil, though it is grown on a wide range of soils ranging from sandy to clay. Soil should be neutral in reaction and well drained. It can be grown successfully on soils having pH range of 6.5 to 7.5

Climate

Pigeonpea is mainly grown in tropical and sub-tropical climate. It is highly susceptible even to light frost. It can tolerate heavy rains provided waterlogging does not take place. It has the capacity to tolerate moisture stress to a great extent because of its deep root system.

Season and varieties of Tamil Nadu

Season	Varieties
Adipattam (June - August)	SA 1, CO 6, CO(RG) 7, Co 8, COPH 2, Vamban 2,VBN (RG) 3, Pusa 84, CO 1, ICPL 87 & 151, Hy 1, 3A & 5, AS 71-37, Bahar, Richa 2000, BRG 1 & 2 (Vegetable), PUSA 991, GAUT 001E
Purattasipattam (September – November)	COPH 2, CO(RG) 7, APK 1, VBN (RG) 3, TVN 3, GAN 1
Summer (February - March)	COPH 2, CO(RG) 7, VBN (RG) 3, Vamban 1, VBN (RG) 3, BSR 1 , Vamban 2, LRG 41, TTB 7

Land preparation

Pigeon pea being a deep-rooted crop respond well to proper tilth. A deep ploughing by soil turning plough followed by 2-3 disc ploughing and harrowing followed by planking is essential. Soil should be levelled well so that water stagnation does not take place. Weeds should be properly removed.

Crop establishment

Spacing

Long duration varieties of pigeonpea are tall, spreading and occupy the field for about 250-270 days. These varieties are planted at wider row spacing of 90-120 cm and about 30 cm between the plants particularly under rainfed conditions. Under irrigated conditions, early maturing varieties are more popular as they fit well in double cropping systems with other crops. These varieties are planted at a row spacing of 50-75 cm and plant to plant spacing of 15-20 cm.

In case of April planted pigeonpea, a row spacing of 90-120 cm is recommended as the vegetative growth is much higher than June planted pigeonpea. Depending upon the size of seed and spacing, 15-25 kg/ha seed of pigeonpea is sufficient.

Sowing

Dibble the seeds adopting the following spacing.

Variety	Pure crop	Mixed crop
CO (Rg) 7	45 cm x 30 cm	120 cm x 30 cm
COPH 2	45 cm x 15 cm	-
Vamban 1, APK 1, Vamban (Rg)3	45 cm x 20 cm	120 cm x 30 cm
CO 6, Vamban 2	90 cm x 30 cm	240 cm x 30 cm
Bund Crop	60 cm for BSR 1 and 30 cm for others	

Sowing time

Pigeonpea is a traditionally a *Kharif* crop sown in June-July with onset of monsoon in various agro-climatic zones of India. Shot duration varieties of pigeonpea are now becoming popular in the irrigated area. These varieties are harvested much earlier than the occurrence of frost/cyclones and can also be fitted well in to multiple cropping systems. Planting of early pigeonpea before the onset of monsoon in the month of June is recommended for higher yields

Planting with onset of monsoon is optimum because the earlier plantings emulate the reproductive phase during the period of heavy rainfall which ultimately caused drop of flowers and pods. Delay in planting reduces the yield of pigeonpea Delay in planting causes reduction in the duration of crop, plant height, number of branches and pods.

Method of sowing

Pigeonpea is generally broadcasted. Line sowing is superior over broadcasting. Broadcasting results in uneven plant population which ultimately results in low yield. In such areas where temporary waterlogging take place, planting on ridges have been found superior.

Seed rate

Seed requirement (kg/ha)

Varieties	CO 6	CO(Rg) 7	Vamban 1	APK 1	Vamban 2	COPH 2	VBN (Rg) 3
Sole Crop	10	25	25	25	10	25	25
Mixed Crop	5	10	10	10	5	-	10

[BSR 1 (Bund planting) 50 g/100 metre]

Select good seeds from pest and disease-free plants.

Post rainy season pigeonpea

When pigeonpea is grown in the cool post rainy season in India, it matures sooner and grows much less tall than when it is sown at the beginning of the rainy season. To obtain satisfactory yields in the post rainy season the crop requires 12-30 plants/m^2, 3-6 times more than required by the same genotype in the rainy season. Since soil moisture often becomes a limiting factor is the pest rainy season, yields tend to declain beyond a certain population. Satisfactory yields can be obtained at 12 plants m^{-2} and any further increase in population led to decline in yield.

Mixed / Intercropping

Redgram is mainly grown as mixed with other traditional crops like urdbean, greengram, groundnut, small millets etc. In this system, seeds of different crops are mixed together and broadcasted.

Redgram can profitably be grown as intercrop with several widely spaced crops like black gram, maize, sugarcane, sorghum, etc., One or two rows of red gram could be sown in between two rows of maize.

Paired planting also can be adopted in redgram where two rows are made closer and thus this extra space is provided in between two pairs of rows. The overall plant population on area basis is not affected but by the adjustment of rows more space is provided in between two pairs which is utilized by intercrop. Summer redgram may be intercropped with spring planted surgarcane.

Ratio in intercropping

Redgram + Jowar / Maize / Pearlmillet = 1:2
Redgram + Greengram / Black gram / Soybean / Groundnut = 1:7

Nutrient management

A quantity of 25:50:25 kg NPK / ha for irrigated crop and 12.5:25:12.5 kg NPK / ha for rainfed condition is recommended

Micro nutrients

Most of pigeonpea cultivars show susceptibility to Zinc deficiency. Foliar spray of 0.5 per cent zinc sulphate with 0.05 per cent lime have to be done for correcting zinc deficiency.

Water management

Pigeonpea is mainly grown in the monsoon season under rainfed conditions. Being a deep-rooted crop, it is capable of extracting moisture from deeper layers in the soil. However, under long period of moisture stress, it may respond to irrigation as well. Two irrigations, one at pre-flowering and other at flowering stage are beneficial in increasing the yield of pigeonpea under rainfed conditions.

Weed management

Two weeding's, one at 25 days and another at 45 days after sowing, are necessary. Among herbicides, Alachlor @ 1.0 kg a.i. / ha is found to be quite effective. Intercropping with short duration legumes like urdbean, mungbean and cowpea is another biological way of controlling weeds in pigeonpea. For short duration crop first 30 days is critical

Cultural methods

Intercropping with sorghum and maize suppresses weeds in long duration crops. With short season crops growing short statured crops such as cowpea, greengram, black gram, groundnut and soybean serve as smoother crops to weed. This not only helps in saving one hand weeding but also gives some additional income to the farmer.

Cropping system

Practice of mixed cropping of pigeon pea with companion crop like sorghum, maize, ragi, black gram, greengram, cowpea and groundnut are very common. Being deep rooted, the crop is very well suited for mixed cropping and enter cropping with the shallow rooted ones. Mixed cropping besides offering an insurance against failure of the crop due to disease, pest, and frost, enables the farmer to obtain variety of crops of their needs from the same piece of land.

The crop generally grown with wide row spacings. However, the genital growth is quite slow and the grand growth starts after 60-70 days of the sowing. A lot of inter-row spaces, therefore, remain vacant during the early stage and get infested by weeds. The space between rows could be profitably utilized by growing short duration crops such as blackgram, moong and cowpea etc. Intercropping with black gram, greengram and cow pea gives about 300-400 kg per hectare additional yield without affecting yield of arhar.

Short duration arhar fits well in following cropping system

- Pigeon pea-wheat
- Pigeon pea -late potato
- Pigeon pea-lentil
- Pigeon pea-sugarcane
- Pigeon pea- wheat moong

Harvesting and storage

Harvest

Green pigeon pea pods are harvested for vegetable. Fully developed, bright green seed is preferred. So, pods should be harvested just before they start losing their green colour. Green pods used as vegetable are commonly picked by hand. Dry seeds of pigeon pea are harvested when the pods are fully ripe and have turned yellow but before the pods start shatter. Nearly mature pods continue to ripen even after plants are cut, but very dry pods shatter and heavy losses occur when plants are cut. Harvesting is usually done manually by using sickle to cut plants and vines, but occasionally by machines and is followed by drying and threshing. Harvested material is dried in the sun in the threshing yard for about a week, depending on the weather conditions. Threshing is done both manually and mechanically. Manual threshing involves beating vines and pods with sticks to separate out the seed, and in some places by cattle trampling. In some places mechanical threshers are used.

Yield: The yield of redgram under irrigated condition is 900 kg / ha and under rainfed it is 500 kg / ha.

PERENNIAL REDGRAM

Variety: BSR 1, Laxmi

Economic uses

Tender beans are pinkish green in colour and can be cooked as curry or added to Kurma or Sabji. When the beans mature, they can be used as dhal. It is recommended for growing in kitchen gardens, backyards, farm road sides, as border crop in sugarcane, banana and betelvine and as a shade crop in turmeric and as a bund crop in paddy double cropped wetlands.

Season: June – July

Height of the plant

150 - 200 cm, Number of branches 7 – 10

Flowering: Five months from date of sowing

Pit Size

Small pits are dug 90 cm apart and the pits are filled with a mixture of well decomposed manure or compost and soil.

Fertilizer application

Urea 15 g and superphosphate 30 g / pit has to be applied.

Planting methods

Two to three seeds are dibbled per pit and watered. When they grow six inches height one plant may be retained in each pit.

Irrigation: Need based

Harvest

Tender pods are harvested for making curry. Each plant will yield two to three kg of green pods at an average seed yield of 750 g to one kg per plant. After the first harvest, the branches are pruned and allowed to grow further. In another 45 – 60 days the plants produce the second flush. For pure crop, about 3 kg of seeds may be required.

Questions

I. Choose the best from the choices given

1. Pigeon pea is originated from
   a) Pakistan b) India
   c) Japan d) Africa
2. The biofertilizer recommended for pigeonpea is
   a) Azospirillum b) Phosphobacteria
   c) Azolla d) Rhizobium
3. The pulse crop having deep root system is
   a) Cowpea b) Blackgram
   c) Pigeon pea d) Bengal Gram
4. The most suitable herbicide for pigeonpea is
   a) Butachlor b) Pendimethalin
   c) 2,4-D d) Alachlor
5. The recommended concentration of DAP for foliar application to pulses is
   a) 2% b) 3%
   c) 4% d) 0.1 %

II. Fill in the blanks

1. __________________ is an indeterminate pulse crop.
2. The herbicide commonly used for pulse crops is______________
3. Water requirement for Arhar is _______ mm.
4. __________ is a perennial redgram.
5. Most of redgram is susceptible to _________ deficiency.

III. Write short notes

1. DAP spray to pulses
2. Perennial redgram
3. Integrated weed management for redgram.
4. Season and varieties of pigeon pea in Tamilnadu
5. Cultural method of weed control in pigeonpea.

10

Greengram, Blackgram & Cowpea-Origin,Geographic Distribution, Economic Importance, Soil and Climatic Requirement, Varieties, Cultural Practices and Yield - Agronomy of Rice Fallow Pulses

Green Gram Mungbean (*Vigna radiata*)

Mungbean is one of the most important pulse crops. It is grown in almost all parts of the country. Mungbean is primarily a crop of rainy season; however, with the development of early maturing varieties, it has proved to be an ideal crop for spring and summer seasons. Middle of March to last week of June is the most suitable growing period where there is a plenty of life-giving sunshine. During this period high temperature and low humidity keep insects and disease infestations at their lowest. Summer season offers an excellent opportunity for raising short duration pulses. Mungbean is an excellent source of high-quality protein. It is consumed in different ways as dal, halwa, snack and so many other preparations. Mungbean is a leguminous crop. It has can fixes atmospheric nitrogen through symbiotic nitrogen fixation. It is also used as green manure crop. Being a short duration crop it also provides an excellent green fodder to the animals. It fits well in various multiple and intercropping systems. After picking of pods, mungbean plants may be used as green fodder or can be incorporated as green manure.

Origin

De Candolle (1986) believes that mungbean has originated in India. According to Vavilov (1926) also mungbean is a native of India and central Asia. It is grown in these areas since pre-historic period. Zukovskij (1962) is of opinion that *Vigna sublobata* which grows in wild from India and Indonesia is the progenitor of mungbean.

Jain and Mehra (1978) reported that *Vigna sublobata* is not the ancestor of mungbean but it appears to be so close with mungbean that some taxonomists have described it as *Vigna radiata* var. *sublobata.*

Distribution

Mungbean is grown throughout the southern Asia including India, Pakistan, Bangladesh, Sri Lanka, Thailand, Cambodia, Vietnam, Indonesia, Malaysia and China, etc. It is also grown in the parts of Africa and U.S.A. and has recently been introduced in Australia.

Area and production

In India, mungbean is grown on an area of about 4.58 million hectares with the production of about 2.61 million tonnes. The average productivity is 547 kg / ha. The major mungbean growing states are Rajasthan, Maharashtra, Andhra Pradesh, Rajasthan, Madhya Pradesh, Bihar, Karnataka, and Uttar Pradesh. Rajasthan stands first in area and production of mungbean in India.

Economic importance

Green gram contains about 25 per cent protein of high digestibility and quality. It is used for various purposes either as a whole or in a variety of ways. Sprouted seeds of green gram synthesize ascorbic acid (Vitamin C) in them. Green gram is also a good source of riboflavin and thiamine. It provides very good quality green fodder to animals. Seed coat, cotyledons and broken parts of embryo after soaking in water can be used as high-quality cattle feed. Being a leguminous crop, green gram fixes 30-40 kg nitrogen per hectare. After picking of pods, green gram plants may be used as green manure. It is also helpful in preventing soil erosion.

Its protein is rich in lysine and poor in S-containing amino acids like methionine and cystine. It is rich in Ca, Fe, K and is a good source of vitamins such as thiamin, niacin and vitamin A. Qualitatively, mungbean has higher digestibility coefficient and biological value. Toxic constituents like tannins (Polyphenols) and trypsin inhibitors are very low in green gram. Hemagglutinins are not present in green gram. It is generally lower in flavus production as they contain relatively less oligosaccharides.

Varieties

Tamil Nadu: CO 1, CO 2, CO 3, CO 4, CO 8, KM 1, ML 131, KM 2, ADT 2, ADT 3, Paiyur 1, Pusa Baisakhi, Mohini (S-8), CO 6, CO (Gg) 7, VBN 2,

VBN(Gg)1, 2 and 3, VBN (Gg) 4, Narender moong 1, Path moong 1, 2 & 3, IPM 205-7, PDM 54, IPU 02-43, LU 391, TU 40, LBG 787, VBN 8.

Soil

Mungbean is grown on a variety of soils ranging from red laterite in southern India to black cotton soils in Madhya Pradesh and sandy soils in Rajasthan. Loam to sandy loam soils is considered ideal for mungbean cultivation. Soil should be well drained. Even temporary waterlogging may damage the crop. Saline-alkali and acidic soils are not suitable for mungbean cultivation.

Climate

Mungbean is grown in summer and *kharif* season in northern India. In southern India, it is also grown in winter season. It requires hot climate and has the capacity to tolerate moisture stress. Areas with annual rainfall of 500-600 mm are considered the best for mungbean cultivation. It can be grown successfully up to an elevation of 2000 meter from mean sea level.

Land preparation

One deep ploughing followed by 2-3 harrowing and planking's are considered optimum for mungbean cultivation. Field should be well levelled and completely free from weeds. In case of summer pulse, sowing is generally delayed till harvesting of main crop. Mungbean could be sown by broadcasting effectively that would not only save time, but would also result in considerable saving in terms of cost of land preparation, irrigation etc.,

Crop establishment

Seed rate

Rainy season mungbean is generally broadcasted with a seed rate of about 15 kg/ha. Sowing of mungbean in rows has been found better as it ensures proper spacing for canopy. A row distance of 30-45 cm should be used for sowing depending upon the time of sowing and varieties. The most proven combination was with a seed rate of 10 kg with 30-45 cm row spacing. Sowing can be done behind the local plough or with seed drill. Seed should be treated with some fungicides before sowing. Treatment of culture should be done after the seed is treated with fungicide. After treatments, seed should not be kept in diffused sunlight. More uptake of nitrogen and increased protein content was possible in wider spacing. High temperatures, bright sunshine hours and low relative humidity where light was probably not a limiting factor, rectangular planting proved better than square planting pattern.

Time of sowing

Kharif

Mungbean is generally sown with the onset of monsoon from June to July. Both early and late plantings have adverse affects on the performance of mungbean. In early sowings crop suffers due to rains at maturity and late sown crop suffered due to poor growth and diseases infestation.

Rabi

In central and southern India, mungbean is sown in *rabi* in the months of October and November. Optimum time of sowing for mungbean is from Oct. 1st and 2nd week. Reduction in yield is possible with advance or delay in sowing

Summer

In North India mungbean is grown in summer season also. The crop is sown in March to early April and harvested in May and June. Only short duration varieties maturing in 60-70 days can be grown with adequate irrigation facilities in this season. Timely sowing of summer mungbean is more important. The delay in sowing affects the yield adversely. The late-sown crop may be affected by the early monsoon rains at harvesting time which may create a problem for harvesting and threshing and ultimately the yield is reduced. If fields are already lying vacant this crop can be sown at any time after mid February. However, if the fields have been cultivated under *rabi* crops, mungbean should be planted immediately after harvesting the preceding crop. In case if the fields are released very late, sowing of mungbean with 'zero tillage' can also be done. Under this practice, mungbean can be sown in the stubbles of wheat just after the latter has been harvested and given irrigation. This practice reduces the cost of cultivation and also saves lot of time which is more important during this season.

Crop rotations

The following are the important rotations

- Mungbean – Wheat / Barley ; Mungbean - Mustard - Urdbean
- Mungbean - Toria - Wheat ; Mungbean - Potato - Onion
- Rice - Wheat - Mungbean ; Maize - Potato - Mungbean
- Maize - Potato - Wheat - Mungbean
- Rice - Sugarcane - (Ratoon) - Mungbean.

Mixed/Intercropping systems

The following are the important mixed / intercropping

Rainy season mungbean is generally sown mixed with several crops like maize, pigeonpea, pearlmillet, sorghum, etc.

Traditionally, seeds of different crops are mixed together and broadcasted without consideration with the plant population and planting geometry.

Intercropping of mungbean in pigeonpea is more profitable than sole pigeonpen. It is also intercropped with cowpea and sorghum.

Nutrient management

a) The following quantity of fertilizers are to be applied basally before sowing.

- Rainfed : 12.5 kg N + 25 kg P_2O_5 + 12.5 kg K_2O +10 kg S*/ha
- Irrigated : 25 kg N + 50 kg P_2O_5 + 25 kg K_2O + 20 kg S*/ha

***Note:** Applied in the form of gypsum if single super phosphate is not applied as a source of phosphorus

Soil application of 25 kg ZnSo4 / ha under irrigated condition is essential.

Water management

Rainy season mungbean does not require irrigation unless there is a prolonged drought. Draining out of extra water from the field is essential for successful cultivation of mungbean. *Rabi* and summer mungbeans may be grown only in the areas where adequate irrigation facilities are available. In general, 3-5 irrigations are required for mungbean in summer season. Increase in moisture supply due to irrigation significantly increases the yield. Giving more than three irrigations does not have any significant impact on yield.

Weed management

One hand weeding after 20-25 days of sowing followed by another after about 20 days may be sufficient.

Application of fluchloralin at 1.0 Kg as pre-emergence 3 DAS + one hand weeding on 25 DAS or pendimethalin at 1.0 Kg/ha 3 DAS + one hand weeding 30 DAS provides excellent control weeds upto harvest. The herbicide Flauzifopbutyl and senthoxydim at 0.75 Kg/ha can also be used for the control of barnyard grass in mungbean.

Multi bloom technology

A special technology is being practiced in Pattukottai block of Tanjore district for blackgram and greengram. The soil is alluvial and rich in organic matter and nutrients. The crop is sown during early summer (Jan.-Feb.) as normal crop and fertilizer is applied as per the recommendation for irrigated crop. In addition to that, top dressing of Nitrogen is done with an extra dose of 25 to 30 kg through urea. Since pulses are indeterminate growth habit and continue to produce new flushes, the top dressing will be done on 40-45 days after sowing. The crop completes its first flushes of matured pods during 60-65th day and put further second new flush within 20-25 days. Therefore, two flushes of pods can be harvested at a time within the duration of 100 days.

Harvest

Green gram should be harvested at its physiological maturity stage. This stage is recognized by uniform shedding of leaves and brown colour of pods with hard seed inside them. Practically, crop is harvested in subsequent stages. First picking is done when pods are moist during morning hours to avoid pod shattering. Subsequent pickings are done at weekly intervals. Varieties with synchronous maturity require only one or two pickings and may even be harvested with sickles. Green crop after picking may be fed to cattle or may be incorporated into the soil. Pods after complete drying under sun are threshed manually. Finally, seeds are winnowed to separate the chaff.

Yield

Under improved package and practices *kharif* green gram may yield 1.2 to 1.5 t/ha, while summer green gram may yield 1.5-2.0 t/ha.

Constraints in production

Green gram has the lowest mean productivity of around 0.38 t/ha on the national basis. The important causes of low green gram, production in India are ecological factors, lack of better plant type and appropriate production and post harvest technology, basic research and socio-economic constraints along with non availability of quality seeds in required amounts. Specific constraints causing low production of green gram may be.

1. Varieties of green gram under cultivation are late maturing and low yielding
2. Yellow Mosaic Virus is a deadly disease of this crop and causes 30 to 70% losses in the grain yield.

3. Stem fly and pod borers cause serious damage to green gram at early and late stages of the crop, respectively.
4. In rainy season, water logging in field for a temporary period causes heavy mortality of plants.
5. Non synchronous maturity of traditional varieties of green gram requires large number of pickings.
6. About 90% of green gram in India is under rainfed conditions. So, crop faces severe moisture stress at grain filling stage, which results in considerable yield reduction

Black Gram Urdbean (*Vigna mungo*)

Vigna mungo is an important *kharif* pulse crop grown throughout India and commonly known as *urd or* mash. It is grown in an area of about 4.60 million hectares in India. Annual production of urdbean in India is about 2.45 million tonnes. It is mainly used as 'dal' and in preparation of many dishes in our diet. The dal of *urd* is rich in protein (24%) and phosphorus acid (385 mg/100g). In southern parts of the country, it is used in preparation of some special dishes. Many South Indian dishes such as *dosa* and *idli* have *urd* as main ingredient. The *urd* flour is also used in making *papad.* Besides, the green fodder of urdbean is very nutritive and is especially useful for milch cattle. It can also be used as green manure. It also acts as cover crop and its deep root system protects the soil from erosion. Urdbean being leguminous has the capacity to fix atmospheric nitrogen and thus helps in restoring the soil fertility. It fixes about 70-90 kg N/ha.

Origin

Urd or blackgram is a native of India and originated from *Phaseolus sublobatus*, a wild plant. There is a mention of urd in vedic texts such as Kautilya's Arthasasthra' and 'Charak Samhita'. Samples of grains have been found at Chalcolithic site and Navdatoli-Maheshwar in India around 1660-1440 B.C.

Distribution

From India it spread to many countries like African, European, American and Asian countries. It has become a popular pulse crop in Pakistan, India, Bangladesh, Burma, Ceylon, and most of the African countries. In India, blackgram is very popularly grown in Andhra Pradesh, Bihar, Madhya Pradesh, Maharashtra, Uttar Pradesh and West Bengal, Punjab, Haryana and Karnataka.

Area and production

In India blackgram is cultivated in an area of 4.60 million hectare with a production of 2.45 million tonnes. The productivity of greengram is 533 kg / ha.

Varieties

Tamil Nadu: ADT 1, CO 1, CO 2, CO 3, CO 4, CO 5, CO 6, KM 1 KM 2, ADT 2, ADT 3, ADT 5, ADT 6, TMV 1, MDU 1, Pant U 30 LBG 17 (Rabi), VBN 3, 4, 5, 6, 7 & 8, APK 1, Panthurd 35, PDU 1, IPU 94-1, K 1, Pusa 1, Pant 430, Khargone 3, ADT 1 to 3, HPU 6, T 65, LBG 402, LBG 22, LBG 20

Soil

It is grown on a variety of soils ranging from a sandy soil to heavy black cotton soils. It thrives well in relatively heavier soil. The most ideal soil would be a well drained loam with a pH range from 4.7 to 7.5. Heavier soils with more water retention capacity are considered ideal for its cultivation. Loam to clay loam soil with neutral pH is the best suited soils for urdbean. It cannot be grown successfully on saline alkali and acidic soils. Urdbean is susceptible to waterlogging. Therefore, soil should be well drained particularly in rainy season.

Climate

It is basically a warm weather crop, chiefly grown during the *kharif* season. However, in recent times in eastern, central and southern India, it is also grown as a winter crop. Black gram tolerates drought and is heat tolerant but is susceptible to frost. It is generally grown in areas which receive about 800 mm rainfall per year. It is not a suitable crop for wet tropics. Urdbean is grown in summer and rainy season in northern India. Heavy and continuous rains at the time of germination and flowering are harmful for the crop.

Season

Kharif urdbean is sown in the last week of June to early July. This crop is mainly grown under rainfed conditions. In case of summer urdbean, sowing may be started in the last week of February till the end of March. Plantings beyond March may be caught by monsoon rains in the second fortnight of June and result in drastic yield reduction.

Seed rate and spacing

During rainy season, the crop is generally sown by broadcasting the seed followed by planking. Sowing could be done in rows with the help of local plough at the distance of 30-45 cm depending upon the variety and sowing time. A seed rate of about 12-15 kg/ha is sufficient depending upon the variety. Urdbean is an important crop in rice fallows in South India. The maintenance of population is very difficult and is the major constraint in rice fallows. 40 kg seed / ha is optimum for urdbean in rice fallows. During summer season, vegetative growth of urdbean is comparatively less, hence higher plant population by reducing the spacing and increasing the seed rate is desirable. Summer crop should be sown at a row distance of 25 cm with a seed rate of 25-30 kg/ha. Sowing can be done behind local plough or with the help of seed-drill.

Seed treatment

Before sowing, seeds should be treated with Carbendazim @ 2.0 g/kg of seed. Seed should also be inoculated with suitable Rhizobium culture, if blackgram is being taken for the first time in the field or after a long duration.

Cropping system

Black gram is generally grown in rotation with following crops

- Black gram – Mustard — North
- Black gram – Wheat — North and North West
- Black gram – Maize – Green gram — North east and east
- Black gram – Wheat — Central parts
- Black gram – Safflower — Southern parts

Blackgram is grown mixed with sorghum, maize, pearl millet and cotton crops during *Kharif* season. The important rotations with black gram are as given below:

- Blackgram - Chickpea ; Maize/Sorghum - Blackgram
- Soybean-Blackgram
- Sesamum/Rainfed Paddy - Blackgram
- Sunnhemp/Pillipesara - Blackgram
- Pigeonpea + Greengram - Blackgram
- Rice - Blackgram

Nutrient management

The fertilizer recommended under rainfed is 12.5 :25: 0 kg NPK / ha while under irrigated condition, it is 25: 50: 0 kg NPK / ha. Under rainfed foliar application of DAP 2% twice once at flowering and 10 days later is recommended.

Water management

The crop cannot tolerate long dry spell and therefore, irrigations are very useful for urd. The crop should be irrigated at flowering and pod filling stages but water should not accumulate in the field. One pre-sowing irrigation at the time of seedbed preparation is required to ensure good germination, if the land is dry. Blackgram is grown purely as a rainfed crop during *Kharif* and on residual moisture in rice-fallows during *rabi*. In case there is a long dry spell the crop is benefited if one irrigation is applied, during the winter season, if irrigation facilities are available, one light irrigation at podding/seed filling stage is beneficial to the crop. Generally, the crop should get irrigation at an interval of 10-15 days. From flowering to pod development stage, there is a need of sufficient moisture in the field. Excess water at any stage affects the crop adversely.

Weed management

When sown as a sole crop 1 or 2 weedings are required in the initial stages to keep the field free from weeds. However, one weeding 30-40 days after sowing is found to be enough. Pre-emergence application of Fluchloralin @ 1 kg a.i/ha in 500 litres of water keeps the field free from weeds for the first 50 days and thereafter the crop smothers the weeds. It should be well incorporated in the soil before sowing. Crop requires two hand weedings at 20 and 45 days after sowing but if hand weeding is not possible, pre-emergence application of Alachlor at the rate of 0.75 kg a.i./ha keeps the field weed free upto 50 days and thereafter the crop smothers the weeds.

Harvest and storage

Blackgram should be harvested when most of the pods turn black. Over maturity may result in shattering. Harvested crop should be dried properly on threshing floor for few days and then threshed. Threshing can be done either manually or by trampling under the feet of bullocks. The grains are separated by winnowing. The grains are stored when moisture contents in grain is 8-10%.

Yield

With improved cultivation practices, the yield will be 1.0-1.2 t / ha grain for *kharif* crop and 0.8 -1.0 t / ha for summer crop.

Constraints in production

Poor soil fertility, inadequate plant population, unavailability of quality seed of high yielding varieties are the main factors contributing to its low productivity. Lack of adequate and timely plant protection measures and failure to conserve optimum moisture further intensifies the problem.

Differences between blackgram and greengram

Characters	Blackgram	Greengram
Stem	Mostly spreading	Mostly erect
Leaves	Mostly yellowish green	Mostly green or dark green
Hairiness	Densely hairy	Sparsely hairy
Pods	Short in length	Long
Seed coat	Black in colour	Green in colour
Hilum	Concave	Flat

COWPEA (Lobia) (Vigna *unguiculata* (L.) Walp.aggreg.)

Cowpea or lobia is a versatile *kharif* pulse because of its smothering nature, drought tolerating characters, soil restoring properties and multipurpose uses. Cowpea is called as vegetable meat due to high amount of protein in grain with better biological value on dry weight basis.

Origin

Vavilov considered India as the Centre of origin of cowpea and Africa and China as secondary centres. But historical and archeological evidences strongly suggested that Africa is most probably the place where the present day cultivated cowpea evolved.

Distribution

It is most widely raised in the tropics and subtropics, particularly in Africa, the Indian subcontinent, southeast Asia and tropical America. Cowpea is mainly grown in Africa. About 90% of the world area is in Africa. In India, it is mainly cultivated in Central and peninsular regions. In northern India it is grown in Uttar Pradesh, Punjab, Delhi and Haryana.

Economic importance

Cowpea grains contain 60 per cent carbohydrates, 22-28 per cent protein, 1.8 per cent fat, 3-4 per cent fibre, 3-3.8 per cent ash and 9-11 per cent moisture. It is a rich source of calcium and iron. The nutritive value of cowpea varies according to climate, soil fertility status and genotype.

The crop forms excellent forage. The feeding value of cowpea forage is high and quite comparable to the Lucerne. It forms a palatable fodder when mixed with cereal fodder. Cowpea gives a heavy vegetative growth and covers the ground so well that it checks the soil erosion in problem areas. Apart from this, being a leguminous crop, cowpea can fix about 70-240 kg/ha of nitrogen per year. On an average, a single season crop under ideal condition can fix about 80-90 kg/ha nitrogen.

Cowpea is often grown as a green manure for soil improvement. Short duration fast growth and low water requirement characteristics of cowpea are well suited for growing it in dry land areas.

Soil

Well drained loams to sandy loam soils are best suited for cowpea. Soil should be neutral in pH. Saline and sodic soils are unfit for cultivation of cowpea. Cowpea prefers deep ploughing for better primary roots development which grows upto 100-150 cm, straight way into the well-drained soil. In normal soils primary roots penetrate upto 50-60 cm, depth. Field should be prepared by giving two or three cross harrowing followed by planking. For summer crop a pre irrigation has to be given immediately after harvesting of *rabi* crop. When the field comes in condition, field should be prepared by giving two or three harrowings.

Climate

Cowpea can be grown in all tropical and subtropical climates. The best temperature for germination of cowpea is 12-15° C. It thrives well between 21 and 35° C. The optimum temperature for primary and secondary root nodulation is 24° C and 33°C, respectively. At higher temperatures hemoglobin content gets reduced. Cowpea is a short-day plant and requires a minimum critical day length of 12.5 hours. For higher production of lobia, 27°C day temperature and 22°C night temperature is required. Areas receiving 600 mm rainfall are suitable for short duration varieties and 600-1000 mm for medium and long duration varieties. It can tolerate drought & therefore, is an important crop of dry areas. It is very susceptible to water logging and is not suitable for heavy rainfall areas. Uniform rainfall distribution is very important for growth and development of cowpea.

Season and varieties

T2, JC 5, JC 10, V 16 (Amba), Cowpea 74, Pusa 152Gomati, Pusa sawami, Gujarat cowpea1 and 2 are commonly cultivated in north India. Kanakamani (PTB1), krishnamani (PTB 2), Pusa 152, V 240, S 288, and S 488 are cultivated in Peninsular India.

Tamil Nadu

Adipattam (June-August): CO 6, VBN 1, VBN 2, VBN 3, CO(CP) 7, Paiyur 1, UPC 622, Grain: C-152, Pusa Phalguni, Amba (V-16) (M), Ramba (V240) (M), Swarna (V- 38) (M), GC-3, Pusa Sampada (V-585), Shreshtha (V-37) (M)

Purattasipattam (September - November): CO 6, VBN 1, VBN 2, CO (CP) 7, Paiyur 1, CO(CP) 7

Summer irrigated: CO 2, CO 6, CO (CP) 7, VBN 2

Field preparation

Land has to be prepared to fine tilth and beds and channels have to be formed.

Seed rate

Optimum plant population of cowpea for one hectare is 4-6 lakhs. However, seed rate of cowpea varies according to the purpose, it is grown. For grain or vegetable 20-25 kg, for fodder 35-45 kg and green manuring 35-40 kg seeds are required.

Varieties	Quantity of seed required (kg/ha)	
	Pure crop	Mixed crop
Paiyur 1, VBN 1, VBN 2, CO 6, CO(CP) 7	25	12.5

Optimum plant population is 3, 50,000/ha.

Seed treatment

Treat the seeds with Carbendazim or Thiram 2 g/kg of seed 24 hours before sowing (or) with talc formulation of *Trichoderma viride* @ 4g/kg of seed (or) *Pseudomonas fluorescens* @ 10 g/kg seed. Biocontrol agents are compatible with biofertilizers. First the seeds should be treated with biocontrol agents and then with Rhizobium. Fungicides and biocontrol agents are not compatible.

Seed treatment with biofertilizer

Fungicide treated seeds should be again treated with a bacterial culture. There should be an interval of at least 24 hours between fungicidal and biofertilizer treatments. The improved rhizobial strain COC 10 is more effective in increasing the yield. The seeds are treated with 3 packets (600 g/ha) of Rhizobial culture COC 10 and 3 packets (600 g/ha) of phosphobacteria developed at TNAU using rice kanji as binder. If the seed treatment is not carried out, 10 packets of Rhizobium (2000 g/ha) and 10 packets (2000 g) of Phosphobacteria with 25 kg of FYM and 25 kg of soil should be applied before sowing. The biofertilizer treated seeds should be dried in shade for 15 minutes before sowing.

Nutrient management

The following is the recommended dose of fertilizers that has to be applied basally before sowing.

- Rainfed: 12.5 kg N + 25 kg P_2O_5 + 12.5 kg K_2O +10 kg S*/ha
- Irrigated: 25 kg N + 50 kg P_2O_5 + 25 kg K_2O + 20 kg S*/ha

***Note:** Applied in the form of gypsum if single super phosphate is not applied as a source of Phosphorus

Soil application of 25 kg ZnSo4/ha under irrigated condition is necessary.

Sowing

The seeds are diddled adopting the following spacing.

Varieties	Sole crop	Mixed crop
CO 6, VBN 1	30 cm x 15 cm	200 cm x 15 cm
CO (CP) 7, VBN 2	45 cm x 15 cm	-
Paiyur 1	30 cm x 15 cm	-

Cowpea is sown by broadcast or line sowing method. The row spacing of 30-45 cm (30 cm for summer and 45 cm for *kharif* cowpea and 10-15 cm of plant spacing are maintained.

In north India, cowpea is grown in *kharif* and spring season. In spring only photo – thermo insensitive varieties are grown. In south India, this crop is also grown in *rabi* season. *Kharif* cowpea should be sown with the monsoon rains. Late sowing affects plant growth adversely which results in poor number of flowers and less grain and pod yield. Summer crop is sown after harvesting of *rabi* crops in the first week of April. Delayed sowing than this date reduces grain yield and sometimes early monsoon rains may destroy the crop. Spring crop in north India is sown in last fortnight of February or first fortnight of March. Early sowing in February results in poor germination due to low temperature and sowing late in March affects grain yield adversely due to forced ripening.

Water management

Irrigation should be given immediately after sowing followed by life irrigation on the third day. Subsequent Irrigations should be given at intervals of 7 to10 days depending upon soil and climatic conditions. Flowering and pod formation stages are critical periods wherein irrigation is a must. Water stagnation should be avoided at all stages. KCl at 0.5 per cent should be applied as foliar spray during vegetative stage if there is moisture stress.

Spraying of diammonium phosphate or urea, NAA and salicylic acid

Foliar spray of Spray of NAA @ 40 mg/litre and Salicylic acid 100 mg/litre once at pre-flowering and another at 15 days thereafter and foliar spray of DAP 20 g/litre or urea 20 g/litre once at flowering and another at 15 days thereafter is recommended.

Weed management

Pre emergence application of Pendimethalin 2 litres 3 days after sowing using Backpack/Knapsack/Rocker sprayer fitted with flat fan nozzle using 500 l of water for spraying one ha. After this, one hand weeding on 30 days after sowing gives weed free environment throughout the crop period. If herbicides are not applied two hand weedings are given on 15 and 30 days after sowing.

Cowpea has a smothering effect on weeds. Because of fast and early growth, it attains full ground cover stage at 30 DAS. Thus, the crop needs protection from weeds for initial 15 to 20 DAS. Two weedings at 15 and 30 days of crop growth ensures an effective weed control. During rainy season, weeds can be controlled by use of chemicals. In sorghum intercropped with cowpea fluchloralin can be applied at 1 kg i.e., / ha in 600 litres of water as pre planting spray. It should be well incorporated in the soil before sowing.

Harvest

Green pods for vegetable purpose can be harvested 45-60 DAS depending upon nature of variety. It is desirable to harvest pods when they are tender and half grown otherwise at later stage they become 'puffy' and develop fibers. The grain crop may mature in 60-100 DAS and harvesting should be done when pods are fully matured. In summer and spring crop, pods should be picked up first and then plants are harvested separately for green forage

Yield

Cowpea yields about 1.2 to 1.5 tonnes of grains and 5.0-6.0 tonnes of haulms per hectare. Vegetable cowpea produces 5.0-6.0 tonnes of green pods per hectare.

Questions

I. Choose the best from the choices given

1. The place of origin of blackgram is
   a) India b) Burma
   c) Japan d) Srilanka
2. The pulse crop which is used as a pulse, fodder and green manure crop is
   a) Mung b) Urd
   c) Cowpea d) Pea
3. Protein of greengram is rich in
   a) Lysine b) Methionine
   c) Cystine d) Trytophane
4. The crop that forms a palatable fodder when mixed with cereal fodder is
   a) Cowpea b) Greengram
   c) Blackgram d) Lentil
5. The herbicide recommended for cowpea is
   a) Pendimethalin b) Fluchloralin
   c) Butachlor d) Lasso

II. Fill in the blanks

1. Critical stage of pulses for irrigation is __________
2. The seed rate of greengram is __________
3. In India, blackgram is grown in an area of __________ m ha.
4. In India production of greengram in ________ m t.
5. The fertilizer requirement of blackgram is __________ NPK kg/ha.

III. Write short notes

1. Multibloom technology
2. Constraints in cowpea cultivation
3. Weed management in pulses
4. Constraints in production of greengram
5. Nutrient management in cowpea

11

Groundnut-Origin, Geographic Distribution, Economic Importance, Soil and Climatic Requirements-Varieties, Cultural Practices and Yield

Oilseed crops are crops which are rich in fatty acid, cultivated for the production of vegetable oil. They are used either for edible or industrial or medicinal purposes. The important oilseeds are groundnut or peanut, sesame or gingelly, sunflower, rapeseed and mustard, soybean, castor, linseed or flax, niger and safflower

Importance of oilseed in Indian economy

In terms of vegetable oils, India is the fourth largest oil economy in the world after USA, China and Brazil.

Oilseeds occupy 14% of country's gross cropped area and contributes to 5% of DNP and 10% of the value of agricultural products. Oilseeds contain 20-60% oil, which is chiefly consumed as food and energy source. They are energy rich cash value crops. Vegetable oils have a number of industrial uses. Oilseeds contain useful carbohydrates essential fatty acids, vitamin A, D, E, K and 18 essential amino acids. Oilseed crops can serve as pasture, cover and green manure crops. They are also used as fodder and for silage. Vegetable oils have medicinal and therapeutic values and also used as laxatives.

Constraints in oilseed production

Oilseeds are energy rich crops, but are grown in energy starved conditions. Most of the oilseed's crops are raised in marginal lands. There is lack of hybrids in oilseeds. Yield losses due to pest and diseases are heavy. Storage facilities are poor. The efficiency of oil extraction units or expellers is very poor. Seed multiplication ratio is low.

Strategies for increasing the oilseed production

More area can be brought under micro irrigation. Good quality seeds should be made available at the right time. Recommended package of practices should be adopted. Field should be kept free from weeds during first 20 – 30 days.

Groundnut (*Arachis hypogeae*)

Origin

The origin is reported to be Brazil. Initial wild sp. *Arachis monticola* a diploid might have given to current tetraploid *Arachis hypogeae*. There is no ancient record on distribution of groundnut to other continents from South America. Spanish, Portuguese, Dutch explorers might have transported it to western pacific, China, India. For India, groundnut is an exotic (foreign) crop. It was imported into India. The early Portuguese navigators appear to have taken the large Brazilian pod forms to Africa and then to India. Similarly, Spanish rulers from Philippines carried the small Peruvian forms to the Philippines, China and eventually to India.

Distribution

Groundnut is cultivated in about 24.7 million ha with a total production of 33 million tonnes in world. India occupies the first place in area and second position in production. In India, it is grown in about 4.89 million ha with a production of 9.25 million tonnes. The cultivation is mostly found in Gujarat, Rajasthan, Karnataka, Tamil Nadu, Maharashtra, Madhya Pradesh and Telangana. It is also grown in Andhra Pradesh, Uttar Pradesh and Punjab.

Among the various states, Gujarat ranks first in total area (17.7 lakh ha) and production (14.4 lakh tonnes). But Tamil Nadu stands first in productivity of groundnut (1981 kg/ha).

Economic importance

Groundnut is an important oil and protein source to a larger portion of population in Asia, Africa and America. Groundnut is grown in the semi-arid zones under fairly low input situation (70 per cent area) and even under such situation the average yield is around 800 to 1000 kg/ha. It is predominantly used for oil extraction in many Asian and African countries. After extraction of oil, the oil cake is used as good manure and cattle feed. The leaf portion of groundnut called ‘haulms’ can be given to the cattle as green fodder. Similarly, the haulms also can be incorporated into the soil which will serve as good organic manure after decomposition. Roasted groundnut kernels are very delicious table dish and rich in protein. Groundnut kernel powder is an important cooking material to prepare various delicious sweets and savories. Similarly, groundnut kernel with jaggery forms a good groundnut sugar candy.

Groundnut seed contains about 45-47 per cent oil and 26 per cent protein. They are rich source of thiamin, riboflavin, nicotinic acid and vitamin E. It also contains phosphorus, calcium, and iron. The oilcake contains 7 to 8 per cent nitrogen, 1.5 per cent phosphorus and 1.5 per cent potash. The crop builds up soil fertility through nitrogen fixation and is an efficient cover crop especially for lands exposed to soil erosion. Because of all these qualities, it is called as "King of oilseeds".

Classification

The cultivated groundnut is divided into other two groups like, Erect or bunch type: *Arachis hypogaea* sub species *fastigiata* and trailing or spreading type: *Arachis hypogaea* sub species *procumbens.*

Climate

Groundnut is grown in continents like America, Africa, near east Asia and Oceania. In southern American continent, the major groundnut growing regions occurs between 20 and 30°S, which receives a rainfall of 1000 and 1400 mm. These are fairly humid regions at an average relative humidity of 73 per cent. Rainy seasons lasts from November to March and the sunshine hours range between 2200 and 2700 per ha. Annual potential evapotranspiration is around 2000 mm and the mean annual temperature is 24°C. Soils are mostly ustic - altisols. Brazil is a major groundnut producing country in this continent.

The diurnal temperature (difference between day and night temperature) affects germination, flowering, pod differentiation and formation. Summer sown groundnut yields more than *kharif* season crop due to ideal climate as clear sky, good light intensity and less incidence of pest and diseases.

Soil

In the Indian sub continent groundnut is a dryland crop grown in alfisols, oxisols or usterts occurring between 7° to 30°N. Heavy and stiff clay soils are not suitable as they interfere with peg penetration into the soil and make the harvest extremely difficult. Red and red loamy and sandy loam soils with neutral pH are the ideal soil types for groundnut cultivation. It is highly susceptible to water logging.

Season

The crop is cultivated in three seasons *viz.*, rainy (*kharif*) season, post rainy season (*rabi*) and summer. *Kharif* season is from June to September while post rainy season starts in November and continue upto February. Average cropping season temperature in India is 27°C.

In Tamil Nadu, the following seasons are practiced for groundnut.

Season	Months
A. Rainfed	
1. Chithiraipattam	(April-May to Aug-Sep)
2. Early Adipattam	(June-July to Oct)
3. Late Adipattam	(July-Aug to November)
4. Aipasipattam	(Oct-Nov to January)
B. Irrigated	
1. Summer	(April - July)
2. Margazhipattam	(Dec-Jan to April)
3. Masipattam region of Tanjavur,	(Feb-March to June)

Groundnut area in Tamil Nadu is 11 lakh hectares with a production of 11 million tonnes.

Varieties

Normally there are five types *viz.*, Bunch, Virginia, Spanish, Valencia and Runner type. However, at present only bunchy type is predominant cultivated type with spreading type cultivation is restricted to small areas. The harvesting problems like retention of more pods in the soil restricted the farmer's choice in selecting spreading type. Some of the varieties under different groups are furnished below.

- ***Bunch type:*** They have light green foliage, broad leaflets and mature early.

  Jyoti, GAUG-1, SG-84, Phule Pragati (JL-24)

- ***Spreading type:*** They have dark green leaves, smaller leaflets and late maturing.

  RS-1, Type-28, Chandra, GAUG-10, Karad-4-11.

- ***Semi spreading type:*** They are intermediate between bunch and spreading types.

  Type-64, Chitra, TG-1, M-197.

The important varieties of Tamil Nadu are TMV 7, TMV 10, TMV 13, TMV 14, Co 3, JL 24, ALR 2, VRI 2, VRI 3, VRI 4, BSR 1, VRI 6, VRI 7, VRI 8, CO

6, CO7, BSR 2, ICGV 00348, ICGV 15083, POL 2, TMV 12, CO1, CO 2, TMV 11, TMV 6, TMV 8

In India

Kadiri-2, Kadiri-3, BG-1, BG-2, Kuber, GAUG-1, GAUG-10, PG-1, T-28, T-64, Chandra, Chitra, Kaushal, Parkash, Amber

Difference between bunchy type and spreading type groundnut

S.No.	Bunchy type	Spreading type
1.	Small to medium sized seed	Medium to large sized seed
2.	Non dormant	Moderately dormant
3.	Plant height upto 60 cm	Height upto 30 cm
4.	Matures in 105 – 120 days	Matures in 130 – 175 days
5.	Pale green leaves	Dark green leaves
6.	Rounded leaflets	Pointed leaflets
7.	Pod distribution is close to the base	Scattered
8.	Testa is rose in colour	Testa is red to purple in colour
9.	Shell is thin	Shell is thick

Field preparation

Groundnut requires a fine loose soil consistency for better setting and development of pods. Usually, groundnut is grown in light soils. Hence the field preparation is easy. Fields are ploughed once with disc plough followed by two tiller ploughing and clods are crushed by disc-harrow. Now-a-days, rotavator is also used for clod crushing and making good tilth in the soil. If ploughing is done with bullocks, iron ploughs are used repeatedly to get fine tilth. Deep ploughing should be avoided because it may encourage pod development in deeper layers. After getting desired tilth, field is thrown into beds and channels, or broad bed furrows of 1.5 m width or ridges and furrows and then seeds are dibbled. This practice is usually followed for irrigated groundnut. In Western African regions, sandy soil does not require land preparation activity, whereas narrow beds cultivation is preferred by the Chinese farmers. In case of bed preparation, 10 m to 20 m^2 beds are formed depending upon the availability of water, soil type and slope.

Seeds and sowing

Quality of seeds

Seeds should be plumpy, hard when pressed and when tasted should not give bad taste. Kernels should be free from rancidity or foul odour. Pods ear marked for seeds should be either hand shelled or shelled through hand decorticator, cleaned. Broken kernels, insect and disease affected kernels should be removed. Kernels with peeled skin should also be discarded because these kernels will not germinate properly. Only hard and full kernels should be retained for seed purpose.

Seed rate

In groundnut cultivation, cost of seeds constitutes 37-40 % of total cost of cultivation. Hence, it is essential to use good quality seeds. A quantity of 140 kg of kernels are used for rainfed condition and 125 kg for irrigated crop. For bold seeded varieties like JL 24, CO 2 and TMV 10, 10 per cent increased seed rate over the recommended seed rate per hectare can be adopted. Seed rate for spreading type is 100 kg/ha.

Spacing

Generally, for bunch types a spacing of 30 - 40 cm x 15 - 20 cm is adopted. For spreading types, a spacing of 45 - 60 cm x 15 - 20 cm is recommended. In Andhra Pradesh, closer spacing of 22.5 x 10 cm is followed for Spanish bunch type.

Under Tamil Nadu condition, a spacing of 30 cm between rows and 10 cm between plants is considered as optimum spacing requirement for groundnut. Wherever groundnut ring mosaic (bud necrosis) is prevalent a spacing of 15 x 15 cm is adopted. A plant population of 3,33,000 plants per hectare is maintained. For spreading type 45 x 10 cm spacing with a plant population of 2,22,000 per hectare is recommended.

Seed treatment

Groundnut responds very well to Rhizobium inoculation. Proper inoculant source and dosage govern the N fixation rate by the nodule bacteria and crop growth response. Rhizobial strains like NC 92, Tal 1000, THA 205 are some of the popular and efficient strains used in all India level. Yield and nitrogen gain response usually exhibited after 3 or 4 cycles of inoculated crops.

In Tamil Nadu, groundnut is treated with Rhizobial culture TNAU 14 developed by TNAU using rice kanji as binder. Seeds are treated with 600 g/ha of rhizobial culture (3 pockets of TNAU 14 Rhizobium culture) and if seed treatment is not carried out, soil application of 2 kg of Rhizobium (10 pockets) mixed with 25 kg of FYM and 25 kg of soil is done just before sowing.

First step: Seeds are treated with thiram or mancozeb at 4 g / kg of seed or carboxin or carbendazim @ 2 g/kg of seed. This seed treatment will protect the seeds from root-rot and collar not infection. Seed treatment can be done 10 or 15 days before sowing.

Second step: Just 30 minutes before sowing, Rhizobium treatment is done and then sowing can be taken up.

If the groundnut seed is treated in the *Trichoderma* @ 4 g / kg or *Pseudomonas fluorescens* @ 10 g/kg of seed (as biocontrol agent), fungicide treatment should be avoided as fungicide and biocontrol agents are not compatible. However, biocontrol agents and biofertilizer are compatible and can be blended at the time of sowing, if needed.

Seed hardening technique

The seeds are graded by using 18/64" round perforated sieve. The graded seeds can be hardened by soaking in 0.5% $CaCl_2$ (50% seed volume) for 6 hours. After 6 hours of soaking, seeds should be taken out and spread over moist gunny bag and covered with another moist gunny bag for 24 hours. Then the seeds with sprouted radicle (first visible expression of radicle) should be separated and dried in shade. It should be repeated by two to three times with 2 hours interval and all the valuable seeds which expressed radicle emergence should be separated and dried under shade. Thus, the viable and dead seeds are separated by this process which enhance planting value. This method ensures 100 per cent population and yield increase by 15 per cent.

Method of sowing

Rainfed condition

Traditionally, groundnut is sown behind the country plough in most of the rainfed area and part of irrigated lands. Three pairs of work animals will cover one acre. Hence, three men labours are required to guide the animals and the plough will open the soil with furrows and women labour will drop the seeds at regular interval in the furrows. Sufficient skill is needed in this method so as to ensure optimum plant population. However, field investigation carried out at Vellore, Tiruvannamalai and Gudiyatham of Tamil Nadu revealed that plant population maintained in this method is far from adequate.

Irrigated condition

In irrigated area, seeds are dibbled in lines at the recommended spacing. On a larger scale seed planter can be used.

Gorru sowing

In chitoor, district of Andhra Pradesh and parts of Vellore gorru sowing is also practiced. Gorru like "Enattu gorru" are used for sowing groundnut and applying fertilizers simultaneously.

Kovai seed drill sowing

A much-improved method of sowing rainfed groundnut is by using Kovai seed drill. This bullock drawn implement has easy draught power (weight 50 kg) with metering mechanism that ensure correct row to row spacing, plant to plant spacing and maintain optimum depth. Kovai seed drill will cover 2.5 acres per day in rainfed area and require one labour for operation. Sowing should be done during June to get higher yield.

Seed depth

An optimum depth of 10 to 15 cm is required to ensure good root growth under rainfed condition while in most of the irrigated area, groundnut in sown at 4 cm depth by hand dibbling. In light soils, seeds are sown deeper than sowing in heavy soils.

Gap filling

To ensure optimum population gap filling is done 10 days after sowing. It should be completed taking advantage of the soil moisture. If the gap filling delayed the yield advantage may not be seen.

Nutrient management

Apply FYM @ 12.5 t/ha as basal. In hard pan red soil, application of 12.5 t of FYM or composted coir pith before chiseling can be done to reduce the hardpan effect.

Groundnut being a leguminous plant requires less nitrogen, more phosphorus and potassium. The following fertilizer schedule is recommended in Tamil Nadu.

Season	N	P	K (kg/ha)
Rainfed	10	10	45
Irrigated	17	34	54

At the time of sowing the recommended fertilizers are applied basally.

Among the micronutrient deficiencies, zinc, boron and sulphur deficiency are common in groundnut. Boron deficiency leads to hollow heart. Micronutrient mixture @ 12.5 kg/ ha mixed with 50 to 60 kg dry sand and broadcasted evenly on the surface of soil.

Calcium deficiency

Symptoms are very rare in leaves and stems, small distorted leaves near branch tips, and terminal buds blacken and fail to continue to develop. Most common symptom in developing fruit is lack of kernel formation, darkened plumule if kernel develops, and reduced seed germination.

Application of gypsum

Gypsum (calcium sulphate) is a very essential nutrient required for groundnut growth and pod yield. Gypsum @ 400 kg is applied per hectare on 40th day of sowing in irrigated crop and in rainfed crop depending upon soil moisture. At the time of gypsum application, soil is thoroughly hoed and earthed up, then gypsum applied and covered. The time of application synchronize with peg formation and pegging period. The peg with fertilized ovule enters easily into the loose soil in gravitational direction and when gypsum applied and incorporated at this time it influences in increasing pod formation and filling of pods. Sulphur present in calcium sulphate also satisfies the requirement of sulphur which is very essential element in oil seeds. Calcium increases the shelling percent and sulphur increases the oil content. Gypsum application should be avoided in calcareous soil as the application will induce creation of hard pan (canker nodules).

Zinc deficiency: A light yellow strip along veins of leaf blade is noticed. Zinc sulfate @ 25 kg per hectare is applied basally. If zinc deficiency is severe foliar spray of 0.5% $ZnSO_4$ is recommended.

Iron deficiency: Interveinal chlorosis is observed. Stunted growth of groundnut is seen. Ferrous sulfate @ 1% is sprayed on 30, 40 and 50 days after sowing.

Earthing up

Earthing up is done simultaneously with second hand weeding at 40 days after sowing with long handled hand hoe. Earthing up will loosen the soil to facilitate peg penetration in the soil and for gypsum application. The soil should not be disturbed after 45 days after sowing.

Water management

Water requirement of groundnut ranges between 400 to 500 mm depending upon the variety and soil. Flowering, pegging and pod formation are the important critical stages for irrigation in groundnut and any moisture stress during these periods will reduce pod yield.

Under Tamil Nadu condition, irrigation scheduling at 0.40 and 0.60 IW/CPE ratio during vegetative and reproductive phases is recommended. About 7 to 9 irrigations are given in Tamil Nadu for both margazhi and chitraipattam groundnut. The last irrigation before harvesting will facilitate the full recovery of pods from the soil.

Micro irrigation: Sprinkler irrigation and micro sprinkler irrigation methods have increased water use efficiency and also enhanced pod yield. Micro sprinkler irrigation scheduled at 100 per cent ETc registered highest pod yield of 3336 kg/ha in flat bed system and under water scarcity area micro sprinkler irrigation scheduled at 67 per cent ETc recorded 3118 kg/ha. Now research on drip irrigation and fertigation in groundnut is going on.

Groundnut crop under rainfed, it does not require irrigation. If dry spell occurs, at least one irrigation should be given at pod development stage.

Weed management

Weeds in groundnut field not only compete for nutrition, moisture and light interception but also hinder easy harvest and sometimes even decrease harvest efficiency. The weed problem is severe in early stage due to its slow growth. The weed competition is critical upto 35 days after sowing. Due to weed competition yield loss upto 70 per cent is found under rainfed condition. Several herbicides are available in market. In several countries, on an average 75 per cent weeds have been reduced by the use of herbicides. In Tamil Nadu, the following weed management options have been recommended.

Pre-sowing application of fluchloralin @ 1.0 kg or pendimethalin 1.0 kg or imazethapyr 50 to 70 g per hectare in 800 to 1000 lit of water and incorporating the herbicide will be effective. The incorporation of herbicide can be achieved by irrigating the field or slight hoeing soon after the pre-sowing application.

Pre-emergence application of fluchloralin 1.0 kg per hectare can be applied followed by irrigation. After 35 to 40 days of sowing, one hand weeding may be given. The other pre-emergence herbicides like metolachlor 1.0 kg or alachlor 0.75 to 1.25 kg or oxyfluorfen 0.25 to 0.50 kg ha can be used along with one hand weeding on 30th day of sowing is more profitable. In case no herbicides application, two hand hoeing and weeding are given on 20 and 40 days after sowing. Fluazifop butyl @ 150 to 250 g per ha as post emergence herbicide on 35 to 40 days after sowing for controlling grasses, especially *Cynodon dactylon.* The soil should not be disturbed through intercultivation after pod formation stage onwards.

Cropping system

Groundnut is a major crop in red soils both under rainfed and irrigated condition. It is grown mixed or intercropped with many crops like pearl millet, maize, sorghum, castor and cotton. The crop is grown in rotation with wheat, lentil, chick pea, barley etc. The most common cropping systems are given below.

- Groundnut – wheat
- Groundnut – barley
- Groundnut – chick pea
- Groundnut – field pea
- Groundnut – lentil

In Andhra Pradesh and Karnataka, sorghum is cultivated after groundnut.

The following systems are very commonly found in Tamil Nadu.

Under rainfed condition in north eastern districts like Vellore and Tiruvannamalai groundnut and red gram (SA1) are grown as intercrops in 6:1 ratio. Within this system sorghum, pearl millet, blackgram, greengram, cowpea and field bean (Mochai) are broadcasted and grown as mixed crop. Groundnut + pearl millet reduces leaf minor incidence in groundnut.

Sequential cropping system adopted in irrigated groundnut is indicated below.

- Groundnut + Gingelly (4:1)
- Groundnut + blackgram (4:1)
- Groundnut+cowpea (6:1)
- Groundnut + cowpea (5:1)
- Groundnut + sunflower (6:2)

Under intercropping system, irrigated groundnut is grown with blackgram, sunflower, gingelly or any other pulses. Cumbu is raised as sequence crop in groundnut (RR) - groundnut (irrigated) crop.

Harvest

The nuts take two months to attain full development. When the older leaves dry and fall on the ground and yellowing of top leaves are observed, it indicates the maturity of groundnut. Plants at random are pulled out and pods are shelled. If the inner wall of the shells is brownish black in colour it indicates that the crop is matured. Prior to harvest, irrigation is scheduled in the field for facilitating easy harvest so that the number of pods retained in the soil will be negligible. If sufficient soil moisture is available at the time of harvest no irrigation is needed. If water is not available, country plough or mould board plough is used to uproot the plants. A labour is engaged to search pods left out in the soil, if necessary. In case of bunch types, the plants are harvested by pulling. In spreading types, harvest is done by spade, local plough or blade harrow or groundnut digger. Pods are stripped from the plants by human labour or by machine stripper developed by TNAU.

Aflatoxins

Groundnut is likely to be contaminated by mycotoxin (toxic metabolites of fungi) called aflatoxins. These are produced by *Aspergillus flavus* and *Aspergillus parasiticus.* They are able to grow under high humidity conditions coupled with high temperature even under low moisture.

Yield

In bunch types, about 1.5 to 2.0 tonnes and in spreading types about 2.0 to 3.0 tonnes of pods per ha can be obtained.

Questions

I. Choose the best from the choices given

1. The origin of groundnut is

   a) China b) Brazil

   c) India d) Japan

2. The seed requirement for bunch type of groundnut is

   a) 60-80 kg kernels /ha b) 80-100 kg kernels /ha

   c) 100-120 kg kernels /ha d) 120-140 kg kernels /ha

3. Gypsum in groundnut is applied

   a) At the time sowing b) At time of podding

   c) At time of pegging d) At time of peak flowering

4. Gynophores of groundnut are called

   a) Pods b) Pegs

   c) Flowers d) None

5. Water requirement of groundnut ranges between

 a) 200-250 mm b) 800-1000 mm

 c) 450-600 mm d) None

II. Fill in the blanks

1. Seed rate of bunch type groundnut varieties is ______
2. Earthing up is done in groundnut crop at _________
3. Spreading types requires _________ quantity of seed rate than bunch type
4. ________ variety of groundnut is having maximum oil content.
5. Shelling per cent for groundnut is __________

III. Write short notes

1. Gypsum application for groundnut
2. Differences between bunch and spreading types of groundnut
3. Groundnut based cropping system
4. Weed management in groundnut
5. Harvesting in groundnut

12

Sesame, Soybean - Origin Geographical Distribution Economic Importance, Soil and Climatic Requirements, Varieties Cultural Practices and Yield

Sesame (*Sesamum indicum* L.)

Sesamum indicum is known as sesame, sesamum, til or gingelly. It is the most ancient crop belongs to the order Tubiflora, family Pedaliaceae.

Origin

The earlier view was that the cultivated sesame originated from Ethiopia. But Bedigian and Harlen (1984) established evidence to indicate that sesame is native of India. In vedic period, liberal mention about Til (Gingelly seed) and Taila (Gingelly oil) are mentioned. The Charred seeds of Gingelly were extracted from Harappa (2500-1500 B.C). These evidences clearly indicate that Gingelly belongs to India. From India, it spread to China and Japan. During 2130 to 2000 BC sesame was valued as currency and seed loans were negotiated just like silver in Persian region. From Mediterranean region it spread to England and Russia. Portuguese introduced sesame to Brazil and it is known as gengelin. The slave trade brought the crop to North America.

Distribution

Sesame is now grown in an area of 6.17 million hectare in Asia, Africa, America with a production of 2.47 million tonnes. The productivity is 404 kg/ha.

The major sesame producing countries are China, India, Myanmar, Sudan, Pakistan, Mexico, Ethiopia, Sri Lanka and Nigeria. The biggest exporters are Sudan and Nigeria and importers are Japan, Italy and Venezuela.

The crop is the most ancient oilseed crop of India next to groundnut and rape seeds and mustard. It is grown in India in about 1.58 million ha with a production of 7.55 lakh tonnes and the productivity is 478 kg / ha. It is cultivated in, Madhya

Pradesh, Uttar Pradesh, West Bengal, Rajasthan, Gujarat, Tamil Nadu and Andhra Pradesh. In Tamil Nadu, it is grown in almost all districts.

Economic importance

The crop has earned a poetic label 'Queen of oil seeds due to high quality of polyunsaturated, stable fatty acid which resists oxidative rancidity. Sesame oil is semidrying oil mainly composed of poly unsaturated oleic acid (18:1) (32 to 54%) and linoleic acid (18:2) (3 to 52 %) with deep yellow colour, pleasant odour and taste. It is considered to be the best table and cooking oil of high stability. The sesame seed contains 40 to 60 per cent oil, 19 to 26 per cent protein and 3 to 6 per cent fiber.

Sesame oil is used as a solvent or carrier for many medicines and cosmetics. A mixture of sesame oil and indigo applied to the skin protects the skin against the heat and cold. Apart from this, the oil seems to have some lubricant effect on the bones of human being. Hence, it is used for 'Oil bath' in Tamil Nadu state.

Sesame oil is also used to repel tsetse fly in East Africa. Besides, the oil is a good substitute for olive oil in salads, pickles and in cooking. The oil is also used in the manufacture of margarine and the poorer grades are used in soaps, paints, lubricants and illuminants. In religious function, in houses, temples the oil is used to illuminate the 'Deepams'.

Sesame seed when fried and cooked is used in soups. When mixed with sugar it forms sweet meat in Africa and Asia. Sesame seed is used in religious ceremonies. Sesame paste (Tahing) is a favoured food in the middle East. Young leaves are used in vegetable soups in West Africa. An instant food and nutrients beverage using 70 per cent soya protein and 30 per cent sesame protein is gaining consumer acceptance. Sesame protein is gaining greater importance.

After extraction of oil, the byproducts formed are oilcake or meal. The cake contains 30 to 50 per cent protein which is rich in sulfur containing essential amino acid - methionine (3 to 4%) but low in lysine. Oil cake is rich in calcium (1 to 2%), phosphorus (0.471 to 1.29%) vitamin C and low in fibre (6 to 8%). The oil cake is an edible cake. It is eaten mixed with sugar by poor people. The cake is a good concentrate feed for animals and poultry. The cake can also be used as manure. It contains about 6.0 to 6.2 per cent nitrogen, 2.0 to 2.2 per cent phosphorus and 1.0 to 1.2 per cent potash.

Classification

1. Based on maturity

a) Early: They have a smaller number of flowers and branches. Ex. Madhavi, Type-12, Type-13, Type-22, JT-7, Tapti, Gujarat til-1, Mirag-1, VRI-1, SVPR-1, TMV-6, Paiyur-1.

b) Late: They have a greater number of flowers and branches. Ex. B-3-1, B-3-2, T-85, Purva-1, B-67.

2. Based on seed colour

a) White seeded: The seeds are white coloured. Ex. TC-25, Haryana til-1, JT-7, Tapti, Mirag-1, SVPR-1, Kanki White, Purva-1, Type-12, Type-13, Type-4, Punjab til-1, Pratap

b) Black seeded: The seeds are brown to black coloured. Ex. Krishna, Type-22, Gouri, Patan-65, Paiyur-1, TMV-6, VRI-1

The promising varieties cultivated in different states are

States	Varieties
Gujarat	Gujarat til-1, Mirag-1, Purva-1, Patan-64, RT-54, RT-54, RT-103, Purva-1, Purba-1, Gujarat til-1, Gujarat til-2, JTS-8, Pragati
Maharashtra	RT-54, JLT-26, Phule til-1, Krishna, AKT-4, MRUG-1, N-128, N-8, Phule til-1, T-85, Tapi, TC-25, RG54, RT-103, JLT-26 (Padma), JTS-8, Pragati
Madhya Pradesh	N-32, JT-7, OMT-11-6-3, Krishna, TKG-55, TKG-22, TKG-21, RT-125, N-32, Kanchan til (JT-7), JTS-8
Uttar Pradesh	T-4, T-12, T-13, T-78, Krishna, T-12, T-13, T-4, T-78, TKG-21, TKG-22, TKG-55, Shekhar (SH446) RT-46, RT-125, JTS-8, Pragati
Bihar	B-14, M-3-1, M-3-3, Kanki White, Krishna, B-67, Krishna, Usha, Uma, TKG-21, TKG-22, TKG-55, RT-125
Rajasthan	Pratap, RT-46, TC-25, T-13, TC-25, RT-46, RT-127, Pratap (C 50), JTS-8, Pragati
Punjab	Punjabi til-1, TC-289, RT-46
West Bengal	OMT-11-6-3, B-67, B-14, B-67, Krishna, Uma, TKG-21, TKG-55, RT-125, Rama, Punjab til-1
Andhra Pradesh	Gouri, Madhavi, T-85, Gauri, Madhavi, Rajeshwasri, Swetha til, Gautam, Varaha, RT-54, RT-103, JTS-1, Chandana (JCS-94), Nirmala (0S-Sel-164), Pragati
Tamil Nadu	CO 1, TMV 3, 4, 5, 6, 7, VRI 3, VRI SV 1, VRI SV 2, SVPR 1, TSS 6, KRR-1, KRR-2, Paiyur-1, VRI-1, VRISV-1, TSS-6, Rama, Nirmala

Climate

Sesame is a tropical and sub-tropical crop. It is a short-day plant requiring 12 hours of sunshine per day. When the bright sunshine hours exceed 13 hours, abnormal growth is seen and flowering is suppressed. Even though warm temperature is required for its growth, sesame comes up very well in winter season. The crop is sown in late monsoon period in dry lands of south Tamil Nadu (November sowing). It is mostly grown in cool temperature and comes up very well by effectively utilizing dew. It is also called as 'Dew crop'. Sesame comes up very well by utilizing residual soil moisture. It can be also grown as catch crop. The optimum temperature ranges between 25°C to 27°C but can do well upto 32°C. At 18°C, germination is delayed, but at 10°C germination is inhibited. Basically, it comes up very well upto 500 m MSL. Low temperature at flowering results in the production of sterile pollen or pre mature flower drop. It is susceptible to water logging and heavy continuous rain. Frost is also harmful to this crop.

Soil

Sesame comes in any type of soil ranging from sandy loam to clay loam or even deep clay which are well drained. The crop thrives the best on sandy loam with adequate soil moisture. The ideal soils will have free drainage with moderate fertility. It comes up very well in neutral soil but also thrives in slightly acidic or saline soil. Extreme saline / alkali soil is not preferable. The pH range should be 5.5 to 8.0.

Seasons

In North India, it is grown in three seasons.

- Rainfed *kharif* crop - June - July sowing (All North Indian states)
- Semi *rabi* crop - Aug- September sowing (Central and Western India)
- Summer crop - January - Feb (Eastern Region)

In South India, the crop is cultivated in *kharif*, *rabi* and summer seasons.

Tamil Nadu

Rainfed

1. Adipattam - (July-Aug)
2. Karthigai pattam - (Nov-Dec)
3. Summer - (April-May)

Irrigated

- Masipattan - (Feb-March)
- Rice fallow - Costal region

Varieties

The promising varieties cultivated in different states are

States	Varieties
Gujarat	Gujarat til-1, Mirag-1, Purva-1, Patan-64, RT-54
Maharashtra	RT-54, JLT-26, Phule til-1, Krishna
Madhya Pradesh	N-32, JT-7, OMT-11-6-3, Krishna
Uttar Pradesh	T-4, T-12, T-13, T-78, Krishna
Bihar	B-14, M-3-1, M-3-3, Kanki White, Krishna
Rajasthan	Pratap, RT-46, TC-25
Punjab	Punjabi til-1, TC-289, RT-46
West Bengal	OMT-11-6-3, B-67, B-14
Andhra Pradesh	Gouri, Madhavi, T-85
Tamil Nadu	CO 1, TMV 3, 4, 5, 6, 7, VRI SV 1, VRI SV 2, SVPR 1

Field preparation

Sesame seeds are very small and hence needs a fine tilth for better germination and establishment. The soil is ploughed with one mould plough, followed by 2 tiller plough and then with disc harrow to obtain crumb structure. If tractor ploughing is not possible, 5-6 ploughing with country plough is given to get fine tilth.

After preparation of field, beds of 10 m^2, 20 m^2 are formed depending upon the availability of water and slope of land. The levelling should be perfect so that there will not be any stagnation of water which will hamper the germination of seeds. In rice fallows, the field is ploughed once at optimum moisture, immediately seeds are broadcasted and the seeds are covered with tiller.

Seeds and sowing

Seed rate

Good quality matured seeds free from pest and fungal damage should be selected. The seed rate depends on the variety, method of seeding and season. The seed rate is 5 to 7 kg / ha for broadcasting and 3 kg / ha for row seeding. In Tamil Nadu condition, a seed rate of 5 kg / ha is required for line sowing. For broadcasting the seed requirement is 9 kg / ha.

Spacing

Under Indian condition, 30 x 10 to 45 x 10 are the recommended spacing followed for gingelly. This will give a plant population of 3,33,000 or 2,22,000 per hectare, respectively. However, in Tamil Nadu, a spacing of 30 x 30 cm is adopted. In rice fallows, the seeds are broadcasted and thinned to maintain 11 plants / m^2.

Seed treatment

Seeds are treated with thiram @ 3 g per kg of seeds or *Trichoderma viride* @ 4 g/kg of seeds. This seed treatment can be done just before sowing. As the bioagents are compatible with biofertilizers, Azospirillum @ 600 g/ha (3 pockets) is used for seed treatment. Such seeds should not be treated with fungicide. For minimizing the bacterial leaf spot, the seeds may be soaked in 0.025 per cent solution of Agrimycin-100 for 30 minutes prior to seeding.

Sowing method

The seeds are very minute. Hence it is mixed with 4 times of its volume with dry sand or powdered farm yard manure and sown evenly along the fertilizer furrows where basal application of fertilizer is done. The seeds are sown at a shallow depth of 3 to 4 cm and covered with the soil. If dibbling or line sowing is not possible, the seeds mixed with sand are broadcasted and covered with tiller. Optimum time of sowing for VRI 1 sesame is 2nd fortnight of February to 1st fortnight of March under summer irrigated condition. But in most situations, it is sown by broadcasting. However, row seeding with seed drills results in higher yield due to better crop stand.

Thinning

Thinning is a very important after cultivation practice which has a significant influence on yield of sesame. Thinning should be done 14 to 15 days after sowing leaving one plant per hill and also maintaining 30 x 10 cm or 45 x 10 cm spacing. If the population is very high, second thinning on 30 days after sowing can be done. If thinning is not done, it will result in lean, lanky plants producing very low yield. About 15 to 20 per cent yield loss has been reported in sesame where thinning is not practiced.

Nutrient management

For Tamil Nadu condition, the fertilizer schedule followed is given below.

	N	P	K (kg/ha)
Irrigated crop	35	23	23
Rainfed crop	23	13	13

The full dose of N, P and K is applied basally for both irrigated and rainfed crop. Along with fertilizer 5 kg of Manganese sulfate is broadcasted at the time of sowing. Shallow furrow application (5 cm soil depth) of fertilizers is done at 30 cm interval and covered with soil. If furrow application is not possible broadcast the fertilizer evenly on the bed before sowing.

Water management

It grows very well in the regions where the rainfall ranges from 500 to 650 mm. At a very minimum level of 300 mm sesame can be successfully cultivated. Flower initiation and capsule filling are the critical stages where there should not be any moisture stress.

Under Tamil Nadu condition, irrigated sesame receives sowing irrigation followed by life irrigation at 7 days after sowing depending upon the soil and climatic conditions. Excessive water is allowed to drain. On 25th day one irrigation is given at pre-flowering stage. Then, one irrigation is given at flowering stage (between 35 to 45th days of sowing) and one or two irrigations at pod setting. During maturity phase (after 65 days of sowing) irrigation should be stopped so that moisture status in the soil is less. If excess irrigation or moisture is retained at maturity phase, it will affect the filling up of the capsules. If there is rainfall during this stage water should be drained. The crop is usually irrigated by check basin or beds and channels method. Irrigation can be scheduled at 50 per cent DASM. Similarly, IW/CPE ratio of 0.6 is optimum. Generally, irrigation is given at an interval of 12 to 15 days. The water requirement of sesame is 350 to 450 mm.

Weed management

In the initial stages sesame is poor competitor of weeds. About 30 to 40 days after sowing are very critical for crop weed competition. During *kharif* season, the crop is heavily infested with weeds. Weeding at the early stages of the crop is essential. Normally weeds are controlled by giving two hoeing and weeding on 15 to 20 days and 35 to 40 days after sowing. Pre planting use of fluchloralion @ 1.0 kg / ha is effective which has to be incorporated in the soil before sowing sesame crop. The promising pre-emergence herbicides are alachlor @ 1.5 kg a.i/ha or fluchloralion @ 1.0 kg / ha. Recent research has shown that alachlor granules @ 2 kg a.i./ha (pre-emergence) plus one hand weeding on 30 DAS is cost effective. There is no need for post emergence application.

Cropping system

Sesame is mostly grown as mixed crop. In North India sesame is grown mixed with pigeonpea, sorghum, pearlmillet, groundnut, cotton and maize. The profitable intercropping systems in India are

- Sesame + soybean (2:1)
- Sesame + greengram (8:2)
- Sesame + pearl millet (1:1)
- Sesame + blackgram (4:2)
- Sesame + Redgram (2:1)
- Sesame + cowpea (8:2)

Sesame is grown as pure crop and it is followed by linseed, chickpea, barley, lentil etc, during *rabi* season.

Harvest

Leaves from the bottom are shed and the top leaves loose their colour and turn yellow at maturity. The colour of the stem turns yellow. The colour of the capsule turns yellow upto the middle. Harvest should be done before the bottom capsule turn brown. Tenth capsule from the bottom is examined by opening and if the seeds turn black, harvest may be taken up for black seeded varieties. The plants are harvested as when they are yellowish brown. The plants should not be allowed to stand dead ripe in the field. Otherwise, there will be considerable loss due to shattering.

Matured plants are pulled out, staked in the bottom one over the other in circle with root and stem pointing out and the seed portion pointing inside. The top is covered with straw so that humidity and temperature is increased. The curing is done for three days. Then the plants are removed one by one and shacked. All the mature seeds will fall. The remaining plants are dried for one more day and again plants shacked. The fallen left out seeds are winnowed, cleaned, dried and stored.

Yield

A well-managed crop can produce about 0.80 to 1.0 tonne of seed per hectare.

Soybean (*Glycine max* L.)

Soybean is cultivated extensively in south Asian countries and U.S.A. This crop was introduced in India in 1977. In our country at present, it is restricted mainly to Madhya Pradesh, Uttar Pradesh Maharashtra and Gujarat. It is also grown on a small area in Himachal Pradesh, Punjab and Delhi. Even though it is the cheapest high quality vegetable protein, we are yet to exploit its full potential. Soybean, being the richest, cheapest and easiest source of best quality proteins and fats and having vast multiplicity of uses as food and industrial products, it is called as Wonder crop.

Origin and spread

The first domestication of soybean has been traced to the eastern half of North China in the eleventh century B.C. or perhaps a bit earlier. Soybean has been one of the five main plant foods of China along with rice, soybeans, wheat, barley and millet. According to early authors, soybean production was localized in China until after the Chinese-Japanese war of 1894-95, when the Japanese began to import soybean oil cake for use as fertilizer. Shipments of soybeans were made to Europe in about 1908, and the soybean attracted worldwide attention. The soybean was a recent introduction to India, probably introduced into India from China, Japan and South East Asia via the Naga Hills and Manipur, at the far eastern tip of India.

Distribution

Soybean is one of the most important crops of the world. It is cultivated in USA, China, Brazil, Mexico and Russia. In India it is cultivated in Madhya Pradesh, Maharashtra, Rajasthan, Himachal Pradesh and Punjab.

Area and production

Soybean is the most popular oilseed in the country after groundnut and soy meal is the largest produced oil meal in the country. In India, the soybean produces 5-7 million tonnes of beans, 1 million tonnes of oil and 3-5 million tonnes of soy meal in a normal year. In India, soybean is cultivated in an area of 10.33 million ha with a production of 10.93 million tonnes. The average productivity is 1058 kg / ha. Madhya Pradesh, Maharashtra, Rajasthan are the major producers of soybean in India. In India 16 parts of whole oil production is soya oil. It is produced 70% in Madhya Pradesh, 18% in Maharashtra, and 9% in Rajasthan.

Economic importance

Mainly the Japanese and Chinese consume it as the Chinese gourmet. Chinese create their delicious works of art with soya. In India people started liking this pulse mainly among the vegetarians. It is a boon to the vegetarians, because it contains the highest protein among the pulses. Other than the whole pulse, lots of processed soya products are available in the market. They include soya milk, soya flour, soya curd and tofu (soya paneer).

Soybean is considered highly nutritive crop. It contains 40 per cent high biological value protein and 20 per cent oil. Soybean protein has 5 per cent lysine, which is deficient in most of the cereals. Enriching cereal flour with soybean improves nutritive quality. Soybean contains less starch; thus, it is good for diabetic patient. Its oil is used as cooking medium and also for making vanaspati ghee. The industrial uses include soymilk, soya flour, soy cake, biscuits, varnish and paints. 1 kg of bean may yield 5-6 kg of soymilk. Soylecithen – a byproduct of oil industry is used as emulsifier in cosmetics and pharmaceuticals. Soybean plant is used as fodder and cake as an excellent concentrate for livestock. It also enriches soil by fixing atmospheric nitrogen. Approximately 85% of soybean produced is utilized for oil extraction, 10% for seed and 5% for food.

Varieties

State	Varieties	Duration (days)
Madhya Pradesh, Uttar Pradesh, Rajasthan	Gaurav (JS-72-280)	102 - 115
Madhya Pradesh, Uttar Pradesh, Rajasthan	JS 75-46, JS 93-05, 95-60, 97-52, PS 1024, PS 1029, Indira Soya 9, MAUS 61, MAUS 61-2	100-106
Madhya Pradesh, Uttar Pradesh, Rajasthan	JS 335, MASU 71, MACS 58, NRC 37, Type 49, Durga, Punjab 1	98-101
Madhya Pradesh, Uttar Pradesh, Rajasthan, Tamil Nadu, Andhra Pradesh	Moneta, TAMS-38, TAMS-98-21	85-85
Arunachal Pradesh, Punjab, Haryana, Uttar Pradesh plains, and West Bihar	PK-416, NRC 2 (Ahilya 1), NRC-12 (Ahilya 2), NRC-7 (Ahilya 3) and NRC-37 (Ahilya 4)	115-120
Arunachal Pradesh, Punjab, Haryana, Uttar Pradesh plains, and West Bihar, Madhya Pradesh, Rajasthan	PK-472, JS 335	100-105
Punjab, Haryana, Delhi, West Bihar,	Pusa-16 DS-7316	105-115
Tamil Nadu	CO 1 (Rice fallow), CO 2, CO (Soy) 3, ADT-1, MACS 124, MACS, 450, Pooja (MAUS 2), Pratikar (MAUS 61) Hardee, Pant Soybean 1029, and Bragg	75-90

Rainfed soybean

Varieties - CO 1, ADT 1, VL Soya –47,

Soil

Soybean can be grown in different types of soils ranging from sandy soil to clay loam soil. The Sandy clay loam, rich in organic matter, is highly suitable for soybean cultivation. It comes up well in neutral pH. The pH of the soil should be 6.0-6.5 so the microorganisms can fix the nitrogen from atmosphere easily and fix it to root nodules. The soil should be well drained because stagnation of water in the field affects the growth of the crop.

Climate

Soybean is sown in the beginning of monsoon and harvested at the end of October. This can be grown in little hot and humid climate. This crop can be grown in both summer and winter. For a good crop of soybean, temperature should be

26.5°C to 30°C or more. For good growth of the plant minimum temperature should be 10°C. When temperature is low, flowering gets delayed.

Sowing

Soybean sowing is done with the first shower and completed in last week of June. Wherever irrigation facility is available, sowing before monsoon gives good results. The row-to-row distance should be 45 cm. When the crop is sown in July, this distance should be reduced to 35 cm. Plant to plant distance should be 4.5 cm and in one acre of land plant population should be 1.5 to 2 lakhs.

Methods of sowing

It is recommended that fertilizer can be applied along with sowing of seed with the help of drill machine. Tractor mounted or bullock drawn seed drill can be used.

If drill is not available, fertilizer should be spreaded evenly and mixed thoroughly to the soil with the help of plough.

Seeds are dibbled at a depth of 2 - 3 cm adopting a spacing of 30 cm x 5 cm. In Erode district, Soybean + Castor (60 cm apart) cropping system gives high net return.

Seed rate

- CO 1 - 80 kg/ha. Optimum plant population 6,66,000/ha.
- CO 2 (irrigated) Pure crop : 60-70 Kg/ha; Inter crop : 25 Kg/ha
- CO (Soy) 3 Pure crop : 50 Kg/ha

Nutrient Management

A fertilizer dose of 20 kg N, 80 kg P_2O_5 and 40 kg K_2O per ha with 40 kg of S as gypsum (220 kg/ha) / ha should be applied as basal dressing. Soil application of 25 kg ZnSo4 / ha is recommended under irrigated condition. Foliar spray of NAA 40 mg/litre and Salicylic acid 100 mg/litre once at pre-flowering and another at 15 days thereafter is essential. Foliar spray of DAP 20 g/litre or urea 20 g/litre once at flowering and another at 15 days thereafter is also recommended.

Water management

Irrigation is given immediately after sowing followed by life irrigation on 3rd day after sowing. Further irrigations at intervals of 7 - 10 and 10 - 15 days during summer and winter season, respectively and the irrigation may be given depending on soil and weather conditions. Soybean is very sensitive to excess moisture and the crop is affected, if water stagnates in the fields. The crop should not suffer due to water stress from flowering to maturity. To alleviate moisture stress, either Kaolin 3% or liquid paraffin at 1% has to be sprayed on the foliage. Water requirement is 450-550 mm. The critical stages are flowering and pod development stage.

Weed management

To control the weeds by chemical, Lasso TM EC – 800 gm or Basalin 400 gm mixed in 300 litres of water and spray before germination. Alachlor may be applied to the irrigated crop at 4 litres/ha or Pendimethalin 3.3 litre ai/ha after sowing followed by one hand weeding on 30 days after sowing. If herbicide spray is not given two hand weedings on 20 and 35 days after sowing may be given.

Weeds of soybean can be controlled effectively by using herbicides as mentioned below:

Herbicide	Rate (Kg/ha)	Mode
Alachlor	1.0 – 2.0	Pre – emergence
Butachlor	2.0	Pre – emergence
Metolachlor	1.5 – 2.0	Pre – emergence
Pendimethalin	1.0	Pre – emergence

Cropping system

Soybean is mainly grown in *Kharif* season. The following is the cropping pattern followed in various states

- Punjab: Soybean-Wheat-Barley
- Haryana: Soybean-Cauliflower-Mustard
- Madhya Pradesh: Soybean-Wheat

Shattering

Dehiscence often reduces soybean yield. Shattering can be high under low humidity conditions. Seed holding ability, despite over ripening is important during mechanized harvesting. When weather conditions are not favourable then harvesting can be performed at 14 to 15% grain moisture. By doing this, the losses occurring during harvesting can be reduced. Generally, there is more loss in early maturing crops during harvesting. If it rains during harvest then harvest should be done immediately after rains. Damage due to rains can be reduced by using all means of drying and harvesting immediately after rains or when rain reduces.

Harvest

When plants mature, leaves turn yellow and starts dropping. Pods dry out quickly.

Yield

Soybean crop can yield around 2.0 to 2.3 t / ha under good management

Rainfed soybean

2. Varieties - CO 1, ADT 1

Season

The crop can be grown in both South-West and North-East monsoon seasons. The middle of July is the optimum time of sowing for rainfed soybean in North Western Zone.

Seed treatment with fungicides and biofertilizers

Seeds can be treated with Carbendazim or Thiram @ 2g/kg of seed 24 hrs. before sowing or with talc formulation of *Trichoderma viride* @ 4 g/kg seed or *Pseudomonas fluorescens* @ 10 g/kg seed. Biocontrol agents are compatible with biofertilizers. First treat the seeds with biocontrol agents and then with Rhizobium.

Nutrient management

The fertilizer should be applied as per soil test recommendation as far as possible. If soil test recommendation is not available adopt blanket recommendation of 20:40:20:20 NPKS kg / ha if adequate moisture is available. Entire dose of N, P, K and S should be applied as basal.

Spacing

The recommended spacing is 30 cm between rows and 5 cm between the plants in the row.

Sowing

Sowing is done by dibbling or by drilling the seeds.

Weed management

If sufficient moisture is available, alachlor may be applied at 4.0 litres/ha or Pendimethalin 3.3 litres/ha after sowing followed by one hand weeding on 30 days after sowing. If herbicide spray is not given, two hand weeding on 20 and 35th day after sowing can be done.

Questions

I. Choose the best from the choices given

1. The recommended planting geometry for sesame is

   a) 25 × 15 cm b) 30 × 10 cm

   c) 45 × 15 cm d) 50 × 20 cm

2. The most ancient oilseed of India is

   a) Sesame b) Groundnut

   c) Sunflower d) Castor

3. The oilseed which is called as queen of oilseeds is

   a) Sesamum b) Groundnut

   c) Sunflower d) Safflower

4. Soybean is originated from

   a) Nigeria b) Tanzania

   c) China d) India

5. The state having the highest production of soybean is

   a) Tamil Nadu b) Gujarat

   c Madhya Pradesh d) Maharastra

II. Fill in the blanks

1. The irrigated soybean is generally spaced at ____
2. _____ oil is considered as poor man's ghee.
3. White sesame has __________ oil content that black varieties.
4. Sesame requires ___________ micro nutrient.
5. Low yield in sesame is due to _________ plant population.

III. Write short notes

1. Nutrient management in Sesame
2. Weed management in Sesame
3. Post-harvest techniques in sesame
4. Economic value of soybean
5. Weed management in soybean

13

Cotton - Origin, Geographic Distribution, Economic Importance Soil and Climatic Requirement, Varieties, Cultural Practices and Yield

COTTON (*Gossypium* spp)

Cotton is often called as "**white gold**". True to its name cotton supports the largest agro based industries in India and cotton industry ranks first in the agro based industries and engages 4 to 5 million people in cotton growing. The industrial sector employs 1.0 million people. At present there are 3660 ginning and 1570 textile mills in India. India earns foreign exchange to a tune of Rs. 2500 million through exports of cotton textiles and products. Cotton is cultivated for 'lint' which is used for preparing threads, yarns and medical cotton mixed with synthetic fibre and for various other purposes. Cotton fibres are used to make nitrocellulose which help to propel solid fuel rockets. With advanced technology, short fibres or fuzzy or linters can now be used to make excellent grade papers like currency papers, linoleum, cellophane, rayon and photographic films and molded plastics, shelter proof glasses, X -ray films. After extraction of cotton fibre, cotton seed crushed in expeller is used as cotton oilseed cake. It has 50 per cent protein 'and used as an excellent concentrate cattle feed and can also be used as manure. Recently cotton seed oil is gaining importance. The cotton seed oil contains 19-22 per cent essential fatty acids like myristic, palmitic, palmitoleic, steric, oleic and linoleic acid. Refined cotton seed oil is an edible vegetable oil and is also used for hydrogenated oil. The crude oil is used for soap making and lubricants.

Fiber

A fiber is defined as a unit of matter with a minimum length of 100 times its diameter, flexible and capable of being woven. The primary natural fibers are from animal sources (wool, silk and hair), vegetable sources (cotton, flax and hemp) and, less commonly, a mineral source (asbestos).

Cotton is a soft fibre that grows around the seeds of the cotton plant. Cotton fibre grows in the seed pod or boll of the cotton plant. Each fibre is a single elongated cell that is flat twisted and ribbon like with a wide inner hollow (lumen). Since cotton fibre has conveltions, it is amenable for spinning

The cellulose is arranged in a way that gives cotton unique properties of strength, durability, and absorbency. It is fresh, crisp, comfortable, absorbent, and flexible, has no pilling problems and has good resistance to alkalis. It has poor wrinkle resistance, shrinkage, poor acid resistance, less abrasion resistance, susceptible to damage by moths and mildew, needs lots of maintenance and stains are difficult to remove. Its fibre length ranges from ½ inch to 2 inches and it has 10% increase in strength when wet.

Cotton is obtained from plant source and it is classified as a natural material as it is obtained from the seeds of cellulose seed fibre staple fibre measuring 10- 65 mm in length and white to beige in color in its natural state. It is composed basically of a substance called cellulose. As cotton occupies 50% of the consumption of fibres by weight in the world, it is called as the king of all fibres.

Bast fibre or skin fibre is fibre collected from the Phloem (the "inner bark" or the skin) or bast surrounding the stem of a certain mainly dicotyledonic plant. The bast fibres have often higher tensile strength than other kinds and are therefore used for textiles, ropes, yarn, paper, composites and burlap. Special properties of bast fibers are that the fiber at that point represents a weak point. They are obtained by the process called retting.

Jute is one of the cheapest natural fibres and is second only to cotton in amount produced and variety of uses. Jute fibres are composed primarily of the plant materials cellulose and lignin. Jute is a long, soft, shiny vegetable fibre that can be spun into coarse, strong threads. It is generally used in geo textiles. It has a good resistance to microorganisms and insects. It has low wet strength, low elongation and is inexpensive to produce

Origin and distribution

Cotton was a semi wild plant often considered as tree plant in pre historic time. Later it was domesticated and four important *spp* of commercial value are found in world. There are four important species of cotton and they are *Gossypium herbaceum* (Uppam cotton) G*ossypium arboreum* (Karunganni cotton) G*ossypium hirsutum* (Cambodia cotton) *Gossypium barbadense* (Egyptian cotton)

G. arboreum spp seem to be originated from India. G. *herbaceum* might have been originated from Arabia, Persia and Baluchistan and introduced into Western India. *Gossypium hirsutum* might have originated from Mexico and *Gossypium*

barbadense would have originated from South America. Major Cotton growing countries are USA, Argentina, Brazil, China, Pakistan, Russia, Uzbekistan, Egypt and Australia.

Area and production

In India cotton is grown in 12.2 million hectares with a production of 37.7 million bales. However, the productivity is around 524 kg / ha. India stands first in area but second in production. USA stands first in production and Australia stands first in productivity. In India, Gujarat, Maharashtra and Telangana are the major states growing around 71 % (8.64 m ha) in area under cotton cultivation and 65% (24.6 m bales) of cotton production.

In India it is grown in 3 zones, viz., 1. Northern zone 2. Central zone 3. Southern zone.

Northern zone comprises of Punjab, Haryana, Northern Rajasthan and Western Uttar Pradesh. It is grown under irrigated condition. It covers 20% of the national cotton area and contributes 40 per cent of the total cotton production. Cotton is grown in alluvial soils. Punjab stands first in productivity of seed cotton (570 kg/ha) followed by Haryana (429 kg/ha)

Central zone encompasses states like Maharashtra, Madhya Pradesh, Chhattisgarh and Gujarat. Central zone covers the largest area in cotton (4.33 million hectares) with production of 4-6 million bales and productivity of 250 kg/ha. Maharashtra is having the largest area in cotton (2.3 million hectare). The crop is mostly grown under rainfed condition (72 %).

Irrigation facilities are available on about 18 % area. This region covers the largest area (55 % of total are) and contribute about 35 per cent in the total production. All the four species are grown. While *hirsutum* and *barbadense* are grown in all states, *arboreum* and *herbaceum* are restricted to Gujarat. It is grown in black soils with undulating topography. Yield levels are low.

Southern zone: Andhra Pradesh, Karnataka and Tamil Nadu are the cotton growing areas in this zone. Tamil Nadu is growing 2.51 lakh hectares with a productivity of 325 kg/ha. All the four species of cotton are being cultivated in this zone.

Hirsutum is the dominant species grown in this region followed by *barbadense* and *arboreum*. More than 68 per cent is under rainfed condition and 32 per cent area is under irrigated condition. This region covers 25 per cent area of national cotton area and contributes 25 per cent of the total seed cotton production. Cotton is grown in red lateritic and black cotton soils in plain as well as in undulating terrain.

Varieties and hybrids (both inter specific and intra specific) are grown here. Besides, cotton is also grown in Bihar, Orissa, Assam, and Tripura in small areas.

In Tamil Nadu it is grown in Coimbatore, Erode, Salem, Trichy, Dharmapuri, Madurai, Theni, Virudhunagar, Thoothukudi, Tirunelveli and Perambalur districts.

Cotton plant exhibits two types of branches viz., Monopodial and sympodial. Monopodial branches are the vegetative branches usually present at the base of main stem and devoid of flowering bud or flower. Sympodial branches are the branches usually arise laterally well above the base and each sympodial branch will bear a square/ flower/ boll. In well-established cotton plant 3-4 monopodial branches and 15-16 sympodial branches can be observed. Hybrids will have more sympodial branches than varieties.

Differences between monopodia and sympodia

S.No.	Monopodia	Sympodia
1.	It is vegetative branch	It is reproductive branch
2.	It will end with leaf	It will end with squre
3.	It will be at acute angle with the main stem	It will be at right angle with the main stem
4.	It will be always in horizontal position	It will be always in vertical position
5.	The number will be one or two per plant	The number will be more than ten per plant
6.	It will be straight	It will be zigzag in appearance

Climate

Basically, cotton thrives very well in hot weather and temperature seems to be a more critical climatic factor which decides cotton production.

For better seed germination a minimum temperature of 15°C is needed. Optimum temperature for seed germination and seedling growth is 32° - 34° C. Vegetative phase may require an optimum temperature of 27-30°C. A very high temperature of well above 40°C affect crop growth. During flowering and fruiting stage an optimum temperature of 27°-32°C is required. Cool night temperature of 15-21 °C is ideal.

Bright sunshine is very important for cotton. A minimum of 4 hours bright sunshine / day is needed. It is grown in the plains as well as upto a height of 1500 m. Low temperature reduces the lint yield. Frost is harmful. As a rainfed crop it is a grown in regions receiving 500 to 2000 mm rainfall. In Tamil Nadu it is grown as rainfed crop in areas where seasonal rainfall is 450-500 mm.

Soil

Cotton is grown in variety of soils having good drainage and it cannot with stand water logging. In India it is grown in alluvial soils (Inceptisols / Entisols) of Northern zone, black cotton soils (vertisols) of southern and western zone, sandy loam to loam (Alfisols) and red lateritic (oxisols) of southern zone

'The best cotton soils are having a texture of sandy clay loam with a depth of 100 cm having a lot of humus (gives black colour) and it is called black cotton soil. It grows in pH of 6.5-8.0. Cotton is fairly tolerant crop in saline soil condition. Acid soils of virgin forest and excessive alkaline soils in arid regions (accumulation of sodium sulphate) are unsuitable for this crop.

Seasons and varieties

S.No	Zone	Districts	Season	Varieties
1.	Winter irrigated (Cambodia tract)	Coimbatore, Erode, Salem,Trichy, Dharmapuri,parts of Madurai, Dindigul, Theni	Aug-Sep to Jan - Feb	MCU5, MCU 9, MUC 13, MCUII, LRA 5166 MCU 12, Suvin Varalakshrni, TCHB213, CO 14, Surabhi, Anjali, Supriya, Suraj, SVPR 1, SVPR 3, SVPR 4, Sumangala, Mallika, Sujata, Suman, LRA 5166
2.	Summer irrigated tract	Madurai, Virdhunagar Tenkasi, Tirunelveli and Cuddalore district	Feb-March	MCU5, MCU8, MCU9, MCU12, MCU13, SVPRI,
			July-Aug	SVPR 2, 3, 4, 5, 6 LRA5I66
3.	a) Winter rainfed zone			
	(Cambodia tract)	Tuticorin, Virudhunagar, Tirunelveli, Sivagangai, Ramnad, Salem, Dharmapuri, Thoothukudi	Middle of Sept to Feb-March	Old varieties MCU 6 (Bharathi), Lakshmi Existing LRA5166, MCU10, New varieties Paiyur I, KC2
	b) Winter rainfed zone			
	(Karunganni tract)	Ottapidaram block in Tuticorin district	Middle of Oct. to Mar-April	Old varieties, K2, 5, 7, 8, 9, K 11, KC2, KC3, K12,
	c) Winter Alfisols rainfed zone	Parts of Madurai	October to Mar-April	New K10, K11, MCU 5
4.	Rice fallow zone	Tanjore, Tiruchi	Jan-Feb to June	MCU 7, ADT 1, SVPR 2, Anjali, CO 15

Hybrids in Tamil Nadu

The important hybrids cultivated in Tamilnadu are CBS 156, Suguna, KCH 1, Savita, HB 224, TCBH 213, Surya, Sruthi

Bt cotton

Bt cotton is an insect-resistant transgenic crop designed to combat the bollworm. Bt cotton was created by genetically altering the cotton genome to express a microbial protein from the bacterium *Bacillus thuringiensis*. It is resistant to bollworm.

Difference between Bt and non Bt cotton

S.No.	Bt Cotton	Non Bt cotton
1.	Total duration is less by 10-15 days	Total duration is more
2.	Resistant to boll worm	Susceptible to boll worm
3.	More than 90% first formed bolls are retained	Only 10-20% bolls are retained
4.	Lint quality is good	Lint quality is poor
5.	More price per kg of seed cotton	Less price per kg of seed cotton

Field preparation

The land is ploughed under ideal moisture condition with either mould board plough or disc plough. This operation is followed by working with cultivator and finally clods are broken either with rotavator or a harrow. Clods are broken to ensure a good seed bed (crumb structure). After breaking clods, ridge plough is used to form ridges at the desired spacing depending upon the variety. The length of the ridge is restricted to a maximum of 10 meters for easy and uniform flow of irrigation water. Either bund or ridges is worked across the ridges at a distance of 10 meters for forming irrigation channels which are rectified manually.

Seeds, seed treatment and sowing

Seeds should be free from pest and diseases and on sowing should show uniform field stand, growth and maturity.

Seed treatment

Cotton fuzz that clings to the seed coat causes difficulty in handling the seeds in sowing. It also harbours pest and disease-causing organisms. The germination of seeds is also affected. Hence delinting is advised.

Cotton seeds are taken in a plastic bucket and treated with concentrated commercial sulphuric acid. For every 1 kg of seed 100 ml acid is added. Seeds are stirred well with wooden stick till the seed become coffee brown in colour. The time taken for this treatment will be between 2-3 minutes depending upon varieties. The seeds are washed 4 or 5 times with fresh water to remove acid. The immature and damaged seeds floating on water surface are removed. The washed seeds are shade dried and treated with fungicide. Acid treatment kills insect larvae, eggs. It also ensures uniform size of the seed which is highly useful for easy sowing:

Three packets (600 g) of Azospirillum biofertilizer is carefully added to rice kanji and to the mixture cotton seeds are mixed thoroughly and kept under shade for 30 minutes. Then the treated seeds are sown in the field immediately.

Seed rate and spacing for cotton varieties and hybrids

Variety/hybrid	Seed rate with fuzz (kg/ha)	Seed rate delinted (kg/ha)	Spacing (cm)	Population / ha
Hirsutum				
MCU4 & MCU 8	25	12.5	75 X 30	44444
MCU 5 & MCU 9	15	7.5	75 X 30	44444
MCU 11	15	7.5	75 X 30	44444
SVPR 1 & SVPR 2	15	7.5	60 X 30	55555
MCU 7 & ADT 1	15	7.5	60 X 30	55555
Barbadense Suvin	-	6.0	90 X 45	24691
Intra specific *Hirsutum* hybrid	-	4.0	90 X 60	18518
Inter specific Hybrid Varalaxmi	-	2.5	120 X 60	13888
TCHB 213	10	-	120 X 60	13888

Depth of sowing

In heavy soils cotton seeds are sown at a depth of 5 cm whereas in lighter soils slightly shallow depth (3 cm) can be adopted.

Skip row planting / sowing

This method of sowing involves sowing 4 rows and skipping 4 rows. This method of sowing is popular in USA and other countries where huge machineries are used. It was reported that skip row sowing gave 30-50 per cent increased yield over conventional planting. The inter alley spacing is used for movement of machinery for doing operation like weeding, hoeing, fertilizer application, plant protection chemical application, harvesting kapas. etc.

Cotton growing system

Cotton geometry is often modified to accommodate short duration crops like pulses, coriander. Normally one row of pulse is used as intercrop. In a wider spacing, two rows of pulses are grown.

Seed rate and spacing for inter crop is furnished below.

Intercrop	Seed rate (kg/ha)	Spacing (cm)	
		Row	Plants
Black gram	12.5	30	10
Green gram	12.5	30	10
Cowpea	7.5	30	20
Soybean	20.0	30	10

Crop rotation

Crop rotation is growing different crops in an orderly sequence on a given piece of land over a period of time. Crop rotation improves the soil fertility; generate continuous employment, increase monetary return, reduce pest, disease, weed incidence, improves the soil structure. In Tamil Nadu, irrigated cotton is rotated with ragi, groundnut, sorghum, maize, chillies, vegetables, sugarcane etc. Some of the sustainable cotton-based rotation is furnished below:

1. Ragi - cotton- sorghum (Coimbatore)
2. Rice - rice fallow cotton (Tiruchi, Tanjore)
3. *Rabi* rice - summer cotton (Srivilliputhur, Rajapalayam) Coimbatore condition
4. Cholam + Cowpea / Green gram / black gram / soybean (Feb-May)
   a) Ragi + Sunflower (as border crop) (May- August)
   b) Cotton + Black gram / green gram / cowpea / soybean (Sep-Jan)
5. Sorghum + Cowpea / green gram / black gram / soybean / (Feb-May)
   a) Onion (May - August)
   b) Cotton (Aug-Jan)

Nutrient management

Winter/ summer irrigated

a) For varieties 60:30:30 NPK kg/ha
b) For Hybrids 120:60:60 NPK kg/ha (Interspecific)
c) 90:45:45 NPK kg/ha (Intraspecific)

Rice Fallow

- For varieties 30:30:30 NPK kg/ha
- Rice stubbles dibbling
- For varieties 30:40:40 NPK kg/ha
- For Hybrids 30:60:60 NPK kg/ha

The entire quantity of P and K and 50 per cent of N are applied as basal to the cotton crop. If basal application could not be done, application can be done on 25th day. For hybrid Jayalakshmi full doses of P and K plus 1/3 N applied basally, at the time of 17-20 days next 1/3 N application on 45th and remaining 1/3 N after 60-65 DAS.

Application of micro nutrients

12.5 kg of micronutrient mixture formulated by the Department of Agriculture, Tamil Nadu is mixed with sufficient sand (50 kg/ha) and broadcasted over the field.

In case of zinc deficient soils, ZnSO4 @ 50 kg/ha should be applied as basal or 0.5 % ZnSO4 spraying thrice after 40 DAS at 15 days interval should be done. The usual secondary nutrient deficiency syndrome is Magnesium deficiency which is often called as "reddening disease in cotton". This deficiency can be corrected by spraying 0.5% MgSO4, urea 1.0 % and ZnSO4 0.1 % as foliar spray on 50 and 80th day after sowing.

Weed management

Cotton is a widely spaced crop and has a slow initial growth. Thus, there is every possibility that weed growth will be very fast in the initial stage of cotton crop (i.e.) upto 60 days of growth of cotton. Hence weed control measures involving cultural, mechanical, chemical methods are to be practiced.

Pre-emergence herbicide application

Pre-emergence application of fluchloralin 2 litres / ha or pendimethalin 3.3 lit/ha is done 3 days after sowing using a hand operated sprayer with deflecting or fan type nozzle. Sufficient moisture should be present. Fluchloralin should be sprayed to the field on the evening and the field should be irrigated immediately. After pre-emergence herbicide application, one hand weeding is done 35-40 days after sowing.

If pre-emergence herbicide is not applied, two hand weeding and hoeing on 18-20 and 35-40 DAS can be given to control weeds.

After cultivation

Being a long duration crop, cotton requires lot of care after sowing. Gap filling, thinning, hoeing and weeding, earthing up and topping are done in cotton. Cotton germination may be completed by 5-7 days. Being a crop of low germination percentage (60 %), gap filling becomes a very important after cultivation operation. Gap filling is done 10th day of sowing. For varieties 3-4 seeds are dibbled wherever gaps are noticed. For hybrids, gaps are filled with already raised seedlings.

Raising hybrid seedling for gap filling

In the case of Jayalakshmi, HB 224, TCHB 213, seedlings are raised in polythene bags of size 15 x 10 cm. Polythene bags are filled with a mixture of FYM and soil in the ratio of 1:3. One seed per bag is dibbled on the same day when sowing is taken up in the field. The seedlings are maintained by pot watering. On 10th day of sowing, poly bag seedlings are planted is each of the gaps in the field by cutting open the polythene bag and planting the seedlings along with the soil intact and then pot watered.

Thinning

This is done 20 days after sowing. One healthy plant is retained and other poor and pest infested plants are removed.

Hoeing and weeding

Hoeing is done first before thinning and 35-45 DAS just before fertilizer application. Junior hoe is an efficient inter cultivator to be worked in cotton. In the operation the ridges are broken without uprooting the plants and country plough is worked is between crop rows after two days.

Earthing up

In the disturbed soil top dressing of fertilizer is done and the soil is earthed up with ridge plough or manual labour and new ridges are formed. Earthing up and covering the basal stem is one of the important cultural practices to reduce the stem weevil incidence in cotton. One or two scrapping of top soil will loosen the top soil and help the cotton crop to grow well.

Topping

Cotton plants tend to grow taller under favourable conditions and put forth more vegetative growth. The excessive vegetative growth should be terminated in order to get more fruiting points and finally bolls. The restriction of vegetative growth is achieved by clipping the terminal and apical buds. This process is known as topping. Topping is done between 70-90 days of cotton crop. Topping at 15th node is done for MCU 9 (long duration cotton) and $10\text{-}12^{th}$ node for MCU 7 (short duration cotton). Topping generally increases the seed cotton yield.

Water management

Cotton requires 550-900 mm of rain for its potential growth. The water requirement depends upon varieties, season, temperature, length of growing season, hours of sunshine, the amount and distribu-tion of rainfall and soil characteristics. In winter Cambodia cotton regions, the water requirement will be around 550 to 700 mm while in summer Cambodia region it will be around 750-900 mm. In the initial stage it can withstand soil moisture stress. After square formation cotton should be sufficiently irrigated. Flower initiation stage, peak flowering and boll development stages are the three important critical stages of cotton in which moisture stress should not be imposed. Moisture stress at peak flowering stage reduces the percentage of fruiting points, boll number. Moisture stress at boll development stage reduces the boll seed index, lint index and hastens early maturity. Excess irrigation causes vegetative growth and hasten boll shedding.

Method of irrigation

Cotton is irrigated by 1) check basin 2) ridges and furrows 3) border strip 4) sprinkler and 5) drip methods. Of all the above methods, irrigation through furrows is commonly practiced in winter Cambodia growing region. Check basin system is gaining importance. The irrigation experiments have indicated that 75 % saving water can be achieved in drip irrigation system without reduction in yield. Besides the weed population is also very low.

Cotton is irrigated at IW /CPE ratio of 0.40 and 0.60 during vegetative and reproductive phase, respectively. After giving sowing and life irrigation to cotton crop, irrigation can be given at an interval of 12-15 days depending upon the soil type. For saving water alternate irrigation or skip furrow irrigation are also followed. In skip furrow irrigations, alternate furrows are skipped permanently when irrigation is scheduled. The skipped furrow is converted into a wider bed and sometimes pulse intercrops are grown in the wider beds. In alternate furrow irrigations, a particular set of alternate furrows is irrigated. During the next irrigation the left-over furrow is irrigated.

Harvest

Harvesting of cotton extends over long period (45-60 days). Seed cotton is picked from the burs. Harvesting is done once in a week and normally 5-7 pickings are taken. Sometimes seed cotton is picked along with bur. This method of harvesting is called cotton snapping. In this method collected bolls are separated.

Picking is done in the morning hours between 10 to 11 AM only when there is no moisture so that dry leaves and bracts do not stick to the kapas and lower the market value. Well and fully burst kapas alone should be picked. Fully burst kapas should be harvested at appropriate time. Any delay in harvest may hamper the quality of kapas due to bad weather. Good quality kapas are sorted out and the insect infected kapas as well as discoloured kapas are separated from the good kapas.

In developed countries, cotton harvesting is done mechanically. For mechanical harvesting defoliation and desiccation are prerequisite.

Defoliation induces abscission (separation) of the cells where the leaves are attached to the plants. Desiccation kills the leaves so that they dry and break off. At least 50 % of the bolls should be open when defoliants are applied. Opening of bolls is speeded up by defoliation. Defoliants are effective at temperature between 80-100°F and they are relatively ineffective below 40°F and above 100°F. (e.g.) Chemoids (boll eye), Paraquat (Gramaxone)

Chemoids, an imported product from USA is sprayed to cotton when 80% boll openings are observed. 2.5 lit of the chemical is sprayed per ha at this stage causing defoliation effect up to 58 % and increasing seed cotton yield up to 20%. Paraquat at 3.5 lit/ha gave 95 % defoliation at 60 % boll opening. However, there was reduction in yield. Another chemical amino triazol is also used in USA.

Post-harvest technology

Immediately after each picking, kapas should be dried in shade. If it is not dried immediately, the colour will change fetching low market price. The kapas should not be dried under direct sun as the fibre strength and luster will be lost. Grade the kapas into good and second quality if it is sorted out at the time of picking. Spread a thin layer of dry sand on the ground and keep the kapas over it. The dry sand will absorb moisture and prevent it from coming into contact with kapas as moisture will stain the kapas and lower quality. Whenever sand spreading is not possible, keep the dry kapas over dried gunny bags or tarpaulin. DO NOT DRY KAPAS OVER FLOOR.

Dried kapas should not be packed lightly in gunny bags. This causes kapas to small fibres and white colour turn into dull brown colour. The store house should be free from rats

Yield

The normal yield of irrigated cotton could be around 1200 to 1500kg/ha.

Management of rainfed cotton

Out of the total cotton area of 2.5 lakh ha in Tamil Nadu, cotton is grown around 1.9 lakh ha under rainfed condition (67-70 % area). In Toothukudi, Virudhunagar and Tirunelveli districts alone rainfed cotton is grown in 75,000 ha which constitute 30 % of total area grown under cotton in Tamil Nadu. The productivity of rainfed cotton is around 210 kg / ha. The productivity of rainfed cotton can be increased up to 1000 kg/ha with better management practices. The management practices are furnished below.

Choice of varieties: LRA 5166, MCU 10 (existing), KC2 (new variety) K 10 and K 11 in Karunganni tract; Paiyur in Dharmapuri.

Seed treatment: Seed hardening with 2 % KCl / DAP for 5 hr. and then shade drying overnight is done to induce seed hardening in cotton to overcome drought in the initial stage.

Seed treatment with biofertilizers: Azospirillum 600 g is mixed with rice kanji and cotton seeds and shade dried for half an hour then sown immediately.

Pre monsoon sowing: Sowing seed hardened cotton seeds two weeks before onset of monsoon is done. The details of probable weeks favourable for pre monsoon sowing in different places are furnished.

Taluks	Pre monsoon sowing time	Corresponding month
Thirumangalam	35-36 standard week	Last week of August
Aruppukottai and Virudhunagar	35-36 standard week	1st fortnight of September
Kovilpatti	39-40th standard week	Last week of September
Kamudhi	39-40th standard week	and 1st week of October

Premonsoon sowing should be done at a depth of 5 cm. This can be achieved if sowing is done by animal drawn gorrus. Kovai seed drill or tractor drawn broad bed furrow cum seeder or tiller drawn seed drill can be used for sowing cotton seeds.

For in situ soil moisture conservation and utilization cotton should be sown either in broad bed furrow system or compartmental bed system.

Cotton + black gram or cotton + cluster beans are ideal cropping system. Black gram varieties APK 1, K 1 and Vamban 1 are suitable for intercropping system and seed rate for intercrops is black gram - 10 kg/ha; cluster bean – 20 kg/ha.

Application of 750 kg of enriched FYM to the main field is very important. It increases the availability of P to the seedlings thereby facilitating better root growth and penetration. The proliferation of root increases the drought tolerance power of cotton as roots can draw water from deeper layers.

Seed rate: The seed rate is 15 kg / ha (fuzzed seed) Note: It is always safe to treat the seed with cow dung when pre monsoon sowing is done manually. Acid delinting can be taken only when machine sowing is adopted. Machine sowing ensures optimum depth.

Spacing: Cambodia cotton- MCU 10, KC 2, LRA 5166; 45 x 30 cm. However, under higher population level (45 x 15 cm) there is significant increase in seed cotton yield. Karunganni cotton - 45 x 15 cm.

Fertilizer recommendation is 40:20:0 N, P, K kg/ha. Entire quantity of N and P is applied if sowing is done immediately after the receipt of first soaking rainfall. In case of pre-monsoon sowing, full dose of P is applied basally and N is applied 20-30 days after establishment, immediately after the receipt of soaking rainfall. It is preferable to apply basally 12.5 kg of micro nutrient mixture thoroughly with 50 kg of sand and broadcasted uniformly.

Weed management: Pre-emergence application of either butachlor 2.5 lit/ha or pendimethalin 3.3 lit/ha or thionbencarb 3.0 lit/ha can be applied as sand mix within 48 hrs. after the receipt of first soaking rainfall for pre monsoon cotton crop. In the monsoon sown cotton, pre-emergence herbicide should be applied within 24 hrs. after sowing. This is followed by one hand hoeing and weeding on 40th day after sowing. At the time of herbicide application sufficient moisture should be present. If pre-emergence herbicide application is not possible, two hand weeding and hoeing on 20 and 40th DAS is done

Gap filling: 3 to 4 seeds are dibbled in places wherever gaps are observed. This is done 7-10 days after germination.

Thinning of seedlings: Thinning of cotton seedlings is done between 15-20th day of sowing leaving one healthy seedling. For pulses thinning is done on 20th day of sowing. While thinning pulses, 15 cm spacing is adopted in between plants

Foliar application

i) Reddening in cotton and its correction: *G. hirsutum*, *G. barbadense* and hybrids manifest reddening of leaves. A combination of Magnesium sulphate

(1 %), urea (1 %) and zinc sulphate (0.5 %) are sprayed as foliar sprays on 45th and 60th day after sowing. The spraying reduces reddening of leave cotton.

ii) In the middle of vegetative phase if there is heavy rainfall, leaching of soil N occurs resulting in N deficiency. To correct this, foliar spraying of urea (0.5 %) is done on 45 and 60th day of sowing which depends on the receipt of heavy rainfall.

iii) Intercultivation with Danti or blade-harrows on 40th day will conserve soil moisture. This should be followed with ploughing in between cotton rows by country plough.

Physiological shedding of buds and bolls

Due to physiological reasons, 60-70% of the fruiting bodies are shed. Major casual factors are environmental conditions, moisture stress, excessive rainfall during reproductive phase, high temperature, water logging at flowering and imbalance in nutrition

Remedial measures

1. Application of NAA 10-20 ppm alternated with DAP 2% can be done.

Leaf reddening

Major reasons for leaf reddening are sudden lowering of night temperature in October and beyond, water logged / higher moisture profile for a long period, increased wind velocity, excessive boll load, lower nitrogen content and deficiency of magnesium

Bad opening of the Bolls

This disorder is known as 'Tirak disease'. The problem is basically concerned with premature and improper cracking of bolls, instead of normal fluffy type of opening. This results in impaired lint and seed quality apart from reduction in yield. The reasons for bad opening of bolls are soil with salinity in the sub soil, light sandy soil, nitrogen deficiency and low humidity during the flowering period.

Remedial measures

The remedial measures are frequent irrigation to retard development of seeds soil salinity or alkalinity, applying the proper correctives for N enrichment and use of growth retardants to check excessive vegetative growth

Most of the recent varieties produce between 100-150 squares, whereas only 22% of these develop into harvestable bolls.

Causes for bad bud and boll development

The causes for poor bud and boll development are: Poor drainage or flooding or inadequate moisture during and after flowering, high humidity during reproductive phase, Zn and Boron deficiency, prolonged drought and moisture stress, high temperature and drought, heavy rainfall during reproductive phase and damage due to pest and diseases.

Cotton fibre quality

Ginning percentage

It is the ratio between lint and seed and is expressed in percentage

$$GP = \frac{\text{Weight of lint}}{\text{Weight of seed cotton}} \times 100$$

Based on the length of fibres, it is classified as follows

1. Short : < 20 mm
2. Medium : 20.5 – 27.0 mm
3. Long : 27.5 – 32.0 mm
4. Extra long : > 32.0 mm

Lint index

It is the weight of lint from 100 seeds

$$\text{Lint index} = \frac{\text{Weight of 100 seeds}}{\text{100-GP}} \times GP$$

Herbicide injury

Cotton is extremely sensitive to 2, 4, D. Even a very small quantity will cause leafy growth of bracts and finger like expansion of young leaves.

Questions

I. Choose the best from the choices given

1. The acid used for delinting cotton seeds is
   a) Sulphuric acid b) Nitric acid
   c) Hydrochloric acid d) Perchloric acid

2. All the four species of cotton can be cultivated in
   a) North zone b) Central Zone
   c) Southern Zone d) Both a and b
3. The state having the highest area under Cotton is
   a) West Bengal b) Gujarat
   c) Bihar d) Maharashtra
4. Defoliation in cotton is important for
   a) Fast growth b) Reduce transpiration
   c) Mechanical harvesting d) Higher yield
5. Among the natural fibres, the fibre amenable for spinning is
   a) Jute b) Cotton
   c) Banana d) Mesta

II. Fill in the blanks

1. Cotton is often called as ___________
2. The dominant species of cotton grown in the southern zone of India is _________.
3. Earthing up helps to reduce _________ incidence in cotton.
4. Reddening of leaves in cotton is due to the deficiency of __________ .
5. In cotton, bad opening of bolls is called as __________ disease.

III. Write short notes

1. Topping in Cotton
2. Intercropping in cotton
3. Post-harvest techniques in cotton
4. Fibre qualities of cotton
5. Acid delinting in cotton

14

Jute - Origin, Geographical Distribution, Economic Importance Soil and Climatic Requirements Varieties, Cultural Practices and Yield

History of jute in India: In Mahabharata times the mention has been made regarding Jute garments as pattagam (patta produced) for jute fibre products and kitajam (insect produced) for silk products. In Ain-e-Akbari, there is mention of sack cloth made from jute, in Rangapur district.

Origin

The primary centre of origin of *C. olitorius* is Africa and the secondary centre may be India or Indo - Burma and South China.

Distribution

Major jute growing countries are India, Pakistan and Bangladesh accounting for 90 per cent of the global production. Other countries of considerable importance are Brazil, Mexico, China, Venezuela, Egypt, Sudan, Sri Lanka, Middle East, Taiwan and parts of tropical Africa and Asia.

Jute (*Corchorus capsularis* and *C. olitorius*) ranks second in importance, next to cotton as a natural fibre and occupies an important place in Indian economy. Jute fibre is extracted from phloem tissue (bast or bark fibre) in the stem of *Corchorus* as against seed fibre in the case of cotton. Of all the textile fibres, jute is the cheapest and is extensively used in the manufacture of packing material for agricultural and industrial products.

Area and production

In India, the area under jute is 6.8 lakh ha and fiber production is 9.3 million bales (1 bale= 180 kg) and the productivity is 2517 kg ha^{-1}.

On an average, West Bengal accounts for about 70 per cent of the area and 75 per cent of the jute production in the country followed by Bihar with 20 per cent of the area and 15 per cent of the production.

Economic impotrance

Jute is used chiefly to make cloth for wrapping bales of raw cotton, and to make sacks and coarse cloth. The fibers are also woven into curtains, chair coverings, carpets, area rugs, hessian cloth, and backing for linoleum.

Jute growing tracts

1. Lower Bengal- The Ganga riverine tract - *Olitorius* tract
2. Melda, Dinajpur tract - *Olitorius* and *Capsularis* tract
3. North Bengal, Brahmaputra valley new alluvium-cooch bihar (WB & Assam)

***Capsularis* tract**

4. Tripura, cachora area of old alluvium - *capsularis* region yield per unit area is very high
5. Koshi area
6. Muzaffarpur, Darbharga Area - *capsularis* area
7. West Bihar, Eastern UP (including Tarai) - *capsularis* area
8. Cuttack - *capsularis* area
9. Orissa

Climate

Humid, tropical (Africa) to warm subtropical (India), High humidity, 300 mm rainfall/month during the vegetative period of about 3 to 5 months after sowing, 100 - 150 cm rainfall is ideal. Young plants cannot with stand water logging. The sowing is adjusted in such a way that the plants attain 1.0 meter height before the heavy and constant rain starts. Ideal temperature: Optimum 24°C - 37°C. A temperature below 20°C is harmful to the crop and it restricts the growth of the plants completely.

Soil

Jute can be grown in all types of soils (except very sandy or heavy clayey). Loam and sandy loam are the best. Extreme acidity / salinity is not suitable for jute. Thrive in old alluvium soils of Bihar, mild acidic soils of Assam, Orissa, light alkaline soils of Tarai district of UP. *Capsularis* thrives in clay loam soils while o*litorius* comes up very well in sandy loam soils.

Varieties

No	Varieties	Season	Duration (days)	Fibre Yield
Corchorus olitorius				
1	Baisake tossa (green stem)	April 15 Uplands	140-150	40 q / ha
2	Chirtali tossa (red stem)	Mar-Apr High lands	140-150	40 q / ha
3.	Bashdow	Mar-May	140-150	40 q / ha
Corchorus capsularis				
1.	Sabuj sona (JRC 212)	High and mid land	120-135	30-35 q/ha
2.	JRC 321	Feb-July Fast growing	110-120	20-25 q/ha
3.	Shyamali (JRC 7447)	High and mid land	120-130	35-40 q/ha

Capsularis: JRC 212, 321, 7447, 4444, 698, 532, 517, 9057, AAUCJ2, KJC 7, JRCM 2, JBC 5, NDC 2008, RRPS 27 C 3, Bidhan pat 1, 2, 3, KTC 1, Padma.

Olitorius: JRO 524, JRO 878, JRO 7835, JRO 632, TJ 40, JRO 3690, KOM 62, JRO 66, JRO 8432, JRO 128, S 19, JRO 204, AAUOJ 1, JBO 2003H, JBO1, JROG 1, JRO 2407, KRO 4, BCCO 6

Preparation of field

Jute requires fine tilth. Usually, 5-6 ploughing followed by planting is necessary for getting the required seed bed.

Nutrient management

It is necessary to apply 4 to 7 t of FYM / compost / ha. *Olitorius* requires less fertilizer than *capsularis*. Application of P increases resistance of plants against lodging and the fibre quality is also improved.

Capsularis

- Rainfed: 40:20:20 kg NPK ha^{-1}
- Supplemental irrigation: 80:40:40 kg NPK ha^{-1}

Olitorius:

- Rainfed: 30:15:15 kg NPK ha^{-1}
- Supplemental Irrigation: 60:30:30 kg NPK ha^{-1}

Sowing

Time sowing

Sowing is done from the end of February for *capsularis* (180-200 days) and beginning of April for *Olitorius* (120-150 days duration). Sowing continues up to June in some areas.

Sowing in low lands during February - Avoid water logging of early Jute crop

Sowing in mid lands - March - April

Upland - May - June

For Bihar, UP - Sowing continued up to mid-July

Method of sowing

Seeds are mixed with loose soil and broadcasted. After broadcast it is covered with soil by planking. By this planking the seeds come into contact with moist soil and seed germination is better. Dropping the seeds in shallow furrows of 3-5 cm deep behind the plough using seed drill and covering the soil is also practiced

Seed rate

Capsularis: For broadcasting 10 kg/ha and for line sowing 8 kg/ha

Olitorius: For broadcasting 6 kg/ha and for line sowing 4 kg/ha

Plant population - 2.5-5.0 lakh / hectare

1. Broadcast: Two thinning are done to maintain proper spacing
2. Seed drill sowing: *Capsularis* - row to row spacing 30 cm; *Olitorius*-20 cm

Intercultivation

Weed crop competition exists up to 60 days of sowing. So, 2-3 weedings at an interval of 20 days after sowing are given. 2-3 hoeing promotes growth.

Water Management

Jute crop requires 500 mm of water. First irrigation is to be given after sowing and life irrigation on fourth day after sowing. Afterwards irrigation can be given once in 15 days.

Cropping system in Jute growing areas

Irrigated: Jute - paddy - potato - wheat

Rainfed

Jute - Paddy

Jute - paddy - pulses - mustard

Harvest

C. Olitorius is harvested at 120-125 days and *Capsularis* at 180-200 days after sowing. Harvesting is done between June to October based on time of sowing and variety used. Early harvesting results in lower yield and weakened fibre. Late harvesting results in higher yield but produce course and inferior quality fibre. The real time of harvesting the crop is when about more than 50 per cent plants are in pods. Sometimes early harvesting is done to accommodate paddy transplanting or to save the jute crop from flood.

Method

Cutting the plants close to the ground is done under dry land condition. When in flooded condition or deep-water situation cutter is driven under water and stem is cut and separated. In low lying UP, Bihar areas where the crop is 50-120 cm under water at harvesting time, the plants are pulled out by hands and the roots are cut off.

In upland condition, the harvested plants are heaped and left in the field for 2-4 days so that the leaves dry up, cuts, rupture, tissues shrink which facilitate leaf shedding, entry of microorganisms into the stem for an early retting when steeped in water. For plants harvested from lowland, steeping is done immediately after harvest.

Steeping

Two to four days after harvest shed the leaves in bundles of about 20-22 cm diameter. Sort the bundle into thin stem and thick stem bundles. Bundle should be kept erect in vertical portion in water for 30 cm depth for 3-4 days before the entire portion is submerged in water (steeping)

For proper steeping the bundles should be laid side by side to form a sort of platform usually in 2 to 3 logs called "Jack" and then submerged in water by putting seasonal logs or concrete slab.

Retting

It is a biological process by which the bast fibre gets loosened for easy separation from the woody stalk. Jack should be submerged for at least 20 cm below the surface of the water, as incomplete submergenc results in under retting and produces 'croppy' fibre of extremely low quality, while over retting results in 'dazed' weak fibre. It is observed that tying of few plants of "Daincha or sunnhemp" in each bundle of jute causes an easy retting.

Gently flowing, fairly deep, clear and soft water are congenial for ideal retting. In case of stagnant water addition of little amount of ammonium sulphate or bone meal especially when atmospheric temperature is low, accelerate retting process. The retting needs around 34°C and retting completes within 10-15 days during July, 30 days during September and afterwards. For finding out extract and point of retting, Jacks must be examined at least once every day after 10-12 days of steeping. When retting is over the fibre comes out after pressing the plant between thumb and finger. Soon after the end points, the jack should be taken out and extraction of fibre must be completed.

Yield

Green plant weight yield is 45 to 50 tonnes per hectare. The fibre yield is 2.0 to 2.5 tonnes per hectare.

Questions

I. Choose the best from the choices given

1. The primary center of origin of *C. olitorius* is
   a) Africa b) China
   c) India d) Bangladesh
2. Jute is mostly sown by
   a) Dibbling b) Broadcasting
   c) Transplanting d) Seed drill
3. The state having the highest area under Jute is
   a) West Bengal b Gujarat
   c) Bihar d) Maharashtra
4. The countries that account for 90 per cent of the global production are.
   a) India b) Pakistan and Bangladesh
   c) Pakistan and Srilanka d) India, Pakistan and Bangladesh
5. The state that accounts for 70 % area and 75 % production of Jute in India is
   a) Bihar b) Orissa
   c) Assam d) West Bengal

II. Fill in the blanks

1. The origin of jute is ____________.
2. *Corchorus olitorious* comes up well in __________ soils.
3. The average yield of *Corchorus olitorious is* _______*q / ha.*
4. The biological process by which the bast fibre gets loosened for easy separation from the woody stalk is called ________
5. The fertilizer requirement of *Corchorus olitorious* under irrigated condition is ___________ kg NPK /ha.

III. Write short notes

1. Harvesting in jute
2. Retting in jute
3. Steeping in jute
4. Jute based cropping system
5. Uses of jute

15

Fodder Sorghum - Origin, Geographic Distribution, Economic Importance Soil and Climatic Requirement Varieties, Cultural Practices and Yield and Fodder Preservation

Forages: The term forages denote plants that are used for feeding the animals that are cultivated in the farm, or it may be a wild used for stock feeding grown for the vegetative purpose. The vegetative phase is encouraged and the reproductive phase is suppressed.

Ideal forage crop for a cropping system

1. It should have more number of leaves.
2. It should have high dry matter production.
3. Leaves should be more succulent
4. It should be nutritive
5. It should have less toxic substances.
6. Stem should be green up to maturity.
7. It should be resistant to pests and diseases.
8. It should have the capacity to ratoon fast.
9. It should be adaptable to a wide range of climatic conditions.
10. It should have more leaf: stem ratio
11. It should have more number of tillers
12. It should be of short duration

    Cereals: Cereals are annual crops rich in soluble carbohydrate and low in protein.

    Grasses: Grasses are either annual or perennial crops rich in carbohydrate and low in protein.

Legumes: Legumes are either annual or perennial crops rich in protein and calcium and low in carbohydrate.

In India, sorghum, maize, cumbu and oats are the major cereal fodders grown both under irrigated and rainfed condition. The first two are common throughout India and oats is popular in North India. The cereal fodders form the main source of energy for the livestock.

FODDER SORGHUM (Jowar) (*Sorghum bicolor*)

It is a favourite fodder in many parts of the country. To improve the nutritive value, it should be grown mixed with leguminous fodder crops like cowpea, cluster bean etc.

Utilization

The crude protein content is 9.2 to 9.8%. Used as green fodder, stover, silage and hay. It is an excellent silage crop. Since it contains HCN, it should be harvested at 50%flowering.

Origin

The centre of origin is Africa. Afterwards it spread to China and India. It comes up well in tropical or sub-tropical climate.

Season

Irrigated

Throughout the year (January-February or April-May is preferred).

Rainfed

June-July or September-October.

Soil

Deep fertile sandy loam soils with good drainage are most suitable for fodder sorghum cultivation. It can also be grown in well-drained heavy soils.

Field preparation

Field is ploughed once with iron plough and twice or more with country plough to obtain good tilth. Ridges and furrows 6 m long and 90 cm apart or beds of 5 x 4 m or 20 m^2 depending on the availability of water and slope of the land are formed.

Nutrient management

Fertilizers are applied as per soil test recommendations. If soil test is not done, the blanket recommendation of 60:40:20 kg NPK / ha for irrigated crop has to be applied as given below.

Irrigated crop

Basal

FYM or compost should be applied at 25 t/ha. Azospirillum can be applied at 10 packets (2000 g). A fertilizer dose of 30: 40: 20 kg/ha of NPK is recommended. Band application of the fertilizer mixture prior to sowing is preferred.

Top dressing

Thirty kg N/ha is applied 30 days after sowing (Band application followed by irrigation).

For the ratoon crop 30 kg N/ha is applied immediately after the harvest of the sown crop and irrigated.

Rainfed crop (as Basal)

A quantity of 12.5 t/ha of FYM or compost can be applied. Nitrogen: 40 kg and Phosphorus: 20 kg/ha should be applied.

Top dressing for the ratoon

Apply 20 kg N/ha after first cut if sufficient moisture is available.

Seeds and Sowing

a) Seed rate: Irrigated: 40 kg/ha: Rainfed: 75 kg/ha

b) Spacing : 30 x 15 cm.

c) Seed Treatment: Azospirillum 3 packets (600 g).

d) Sowing: Sowing is done to a depth of 3 cm on the side of the ridge and covered with soil. Sowing using seed drill or broad casting is also practiced.

Varieties

Suitable varieties for irrigated (Jan - Feb and Apr - May) are Co.11, Co. 27, Co.F.S. 29

Suitable varieties for rainfed conditions (Jun - Jul) are Co.11, Co27, Co.F.S.29

Suitable varieties for rainfed (Sep - Oct) are K7, Co.27, Co.F.S. 29, K 10

Other varieties are SSG 59-3, Ruchira Maldandi, PCH 106, Jawahar Chari-69, Proagro Chari (SSG -988), FSH-92079, CSH 20MF, S-513.

After cultivation

Weeding on 20th day along with thinning and gap filling is done to maintain spacing and there by population. Subsequently one hand weeding may be given on 35-40 days after sowing.

Irrigation

Irrigation immediately after sowing and life irrigation on the 3rd day and thereafter once in 10 days are being recommended.

Harvest

Harvesting is done 60-65 days after sowing (50% flowering stage).

Green Fodder Yield

First cut: 45 t/ha and Second cut: 25 t/ha.

Note: Mixed cropping with cowpea in the ratio of 1:1 sorghum: cowpea will help to improve fodder yield, quality of fodder and soil fertility. Under rainfed conditions calopo or stylo may be a suitable mixture.

Sorghum poisoning

Young plants of sorghum (30-40 days stage) contain cynogenic glucoside "Dhurrin". Dhurrin in the stomach of animals is converted into hydrocyanic acid. Thus, when young plants (about 5 kg) of sorghum are fed to livestock, it causes death of the animal. This is known as "prussic acid poisoning" or "sorghum poisoning". HCN content is more in leaves. Concentration of HCN is more in the morning and in summer.

The HCN content gets diluted as the age advances. Young leaves contain more. High N applied crop contains more. When high N with poor P is applied, it also produces HCN. Prussic acid content in leaf is 3-25 times greater than stalks. Toxic level of prussic acid is >200ppm. After flowering / heading, the content reduces to safe level for feeding. Individual animals differ in the ability to resist the poisoning. To overcome this problem, feeding of sorghum green fodder before 50 days stage should be avoided.

MULTICUT FODDER SORGHUM

Sorghum is one of the widely adopted forage crops due to its high yielding ability, better nutritive value and suitability for ratooning. It occupies maximum area among different fodder crops. Sorghum fodder is suitable for silage and hay making. In India, fodder sorghum is grown in 2.6 m ha mainly in western UP, Haryana, Punjab, Rajasthan and Delhi and fulfills over two third of the fodder demand during Kharif season. Forage sorghum plant grows 6 to 12 ft tall and produces more dry matter tonnage than grain sorghum. Sorghum is fast-growing, warm weather annual that can provide plenty of feed in midsummer during lean period. Sorghum is best suited to warm, fertile soils whereas cool, wet soils limit its growth. The crop tolerates drought relatively well, though adequate fertility and soil moisture maximize sorghum yields. The plant becomes dormant in the absence of adequate water, but it does not wilt readily. Growth resumes when moisture conditions improve. Multicut fodder sorghum is more advantageous in many ways such as high yield in short period, saving in terms of seed and land preparation.

Variety

The important varieties of multicut fodder sorghum are TNFS 0952 (192 t/ha/yr), CO 27 (rainfed fodder), CO (FS) 29 (167 t/ha/yr) and CO 31 (Multicut Cholam) (190 t/ha/yr), CSV 30 F, CSV 32 F, CSV 33MF (225 t/ha/yr). As the crop has more leaves and the stem is highly succulent in nature, the green fodder is highly relished by cattle. It contains high protein and less crude fibre and hence higher digestibility

Season

Sorghum can be grown throughout the year as multicut variety under irrigated conditions. This crop requires an annual rainfall of 300-400 mm.

Soil

All types of soil with good drainage. It does not come up well on heavy clay soil or flooded or waterlogged conditions. However, too much sandy soils should be avoided.

Preparatory cultivation

The field has to be ploughed 2-3 times to obtain a good tilth. Ridges and furrows of 6 m long and 60 cm apart have to be formed. Farmyard manure at 25 tonnes / ha should be applied before ploughing and incorporated well.

Seed rate

The seed rate is 5 kg / ha

Spacing

- The spacing is 30 x 15 cm (Sowing should be done on both sides of ridges)
- For seed production, the spacing is 60 cm x 15 cm
- Seeds have to be treated with 3 packets (600 g)/ha of Azospirillum and 3 packets (600g) of Phosphobacteria or Azophos 6 packets (1200g)

Nutrient management

- The crop has to be supplied with well decomposed FYM @ 10 t ha^{-1} three weeks prior to sowing of the crop.
- The recommended dose of fertilizer to be applied as basal is 45: 40: 40 kg NPK/ha
- A quantity of 45 kg N should be top dressed 30 days after sowing
- After each harvest, 45 kg N/ha should be applied.
- After first year 45:40:40 kg NPK/ha has to be applied.

After cultivation

First weeding should be done on 25-30 days after sowing. After each harvest a weeding may be given before fertilizer application.

Irrigation

Irrigation should be given once in 7-10 days depending upon soil condition

Plant protection

Plant protection is generally not needed. If shoot fly is noticed, either Endosulphan 35 EC 500 ml/ha or Dimethoate 30 EC 500 ml/ha in 250 litres water may be sprayed. Plant protection sprays may be stopped one month before harvest of green fodder.

Harvest

Harvesting for green forage can be done when the crop has attained 50 per cent flowering. First cutting can be made 70 days after sowing and subsequent cuttings at an interval of 50 days. The plants are cut 5 cm above the ground level at the time of each cutting.

Fodder

First harvest 65-70 days after sowing and there after the ratoon crop may be harvested once in 50 days depending on flowering.

Seed

The crop for seed is harvested 110 – 125 days after sowing

Yield

The yield of green fodder is 192 t / ha / year in 6-7 harvests: Seed yield is 1000 Kg / ha / year. Seeds can be harvested thrice in a year. Fresh seeds have dormancy for a period of 45-60 days and hence should be used for sowing only after 60 days

Storage

Can be fed to cattle as green fodder or dry fodder and also ensiled.

Questions

I. Choose the best from the choices given

1. The recommended spacing for irrigated fodder sorghum is

   a) 25 × 15 cm   b) 30 × 10 cm

   c) 45 × 15 cm   d) 30 × 15 cm

2. The seed rate for irrigated fodder sorghum is

   a) 20 kg / ha   b) 30 kg / ha

   c) 40 kg / ha   d) 50 kg / ha

3. The fodder yield of irrigated fodder sorghum is

   a) 25 t / ha   b) 35 t / ha

   c) 45 t / ha   d) 55 t / ha

4. Sorghum fodder is a rich source of

   a) Vitamins   b) Protein

   c) Minerals   d) Soluble carbohydrate

5. In multicut sorghum fodder crop, first harvest can be done at

   a) 30 DAS   b) 60 DAS

   c) 90 DAS   d) 120 DAS

II. Fill in the blanks

1. Sorghum crop has originated from ________.
2. The spacing for irrigated fodder sorghum is ________.
3. Young plants of sorghum contain ________ which is poisonous to livestock.
4. For the ratoon crop ________ kg N / ha is applied immediately after the harvest of the main crop
5. The seed rate for fodder sorghum under rainfed condition is ________ kg / ha.

III. Write short notes

1. Sorghum toxicity
2. Fertilizer application to fodder sorghum
3. Varieties of multicut fodder sorghum
4. Nutrient management in multicut fodder sorghum
5. Weed management in multicut fodder sorghum

16

Napier-Bajra Hybrids - Origin Geographic Distribution, Economic Importance, Soil and Climatic Requirement, Varieties, Cultural Practices and Yield

Fodder Grasses

The term fodder grasses denote grasses that are cultivated in the farm and used for feeding the livestock.

NAPIER-BAJRA HYBRIDS *Pennisetum glaucum* and *P. purpureum*

Cumbu Napier hybrid grass is a cross between Cumbu (*Pennisetum glaucum*) and Napier grass (*P. purpureum* Schumach.), widely cultivated across India, Africa, Sri Lanka and South East Asian countries. As the hybrid is a triploid, it displays complete sterility and high vegetative growth. In Tamil Nadu, land area utilized for growing fodder is negligible, accounting only 1.6 per cent of the total cultivated area

Napier-bajra hybrids often also referred as Bajra-Napier hybrids are tall growing (200-300 cm), erect, stout, deep rooted, perennial hybrid grass derived by inter specific hybrid between *Pennisetum glaucum* and *P. purpureum*. The hybrid is a triploid and hence sterile. They spread by short stout rhizome to form large clumps or stools upto 1 meter across. N-B hybrids grow well on deep, retentive soils of moderate to fairly heavy texture and also grow on light textured soils, but sandy loam soil to loam is preferred for better response to management. These are propagated by two nodded stem cuttings or by division of root stock. Among the grasses, N-B hybrids are the best green forage yielders in a unit time and space.

Origin

Napier grass is native of Rhodesia in South Africa, where it is found growing extensively.

Distribution

It is widely distributed in tropical and sub-tropical regions of Asia, Africa, southern Europe and America. In India, it is cultivated in Punjab, Haryana, Uttar Pradesh, Bihar, Madhya Pradesh, Odisha, Gujarat, West Bengal, Assam, Andhra Pradesh and Tamil Nadu.

Economic importance

It is more vigorous, nutritious, succulent, palatable and responds to heavy nitrogenous fertilization. Young leafy fodder is highly palatable and of fairly good quality, but mature grass has large proportion of stem. When cut 10-15 cm from the ground at regular intervals of 45 days, it yields 300-350 t green fodder / ha.

Season

Irrigated crop can be planted throughout the year.

Soil

Loamy soil with good drainage is suitable. It also comes up well in acidic soils.

Field preparation

Field is ploughed once with iron plough and twice with country plough. Ridges and furrows 6 m long and 50 cm apart are formed along with suitable irrigation channels.

Varieties of N-B hybrids

S.No.	Varieties	Area of cultivation
1.	Hybrid Napier No.3	Andhra Pradesh, Himachal Pradesh and Assam
2.	Pusa giant and NB-21	Whole of the India and tropics
3.	Co 3, Co (CN) 4, Co (BN) 5	Tamil Nadu, Karnataka, Andhra Pradesh and Gujarat
4.	IGFRI-3 and -6	Central India, north-east hills and Northern hills
5.	IGFRI-7	Hilly, sub-humid and sub-temperate regions of India
6.	PBN-83	Punjab
7.	IGFRI-10	Whole country
8.	Heshwant (RBN-9)	Maharashtra

Varieties for Tamilnadu

The varieties CO3, CO (CN)4 and CO (BN)5 are the important varieties for Tamilnadu.

Special features of CO (BN) 5

It gives high green fodder and dry matter yield. The leaves are broader and stems are succulent. Winter hardiness and gives stable yield throughout the year. It exhibits quick regeneration capacity.

Manuring

Fertilizers are applied as per soil test recommendation along with recommended FYM / compost. If soil test is not done, the blanket recommendation of 150:50:40 kg NPK / ha is followed.

a) **Basal**

FYM or compost should be applied at 25 t/ha. It is recommended to apply 10 packets (2000 g) of Azospirillum. A fertilizer dose of 75: 50: 40 NPK kg/ha should be applied.

Band application of the fertilizer mixture prior to planting is preferred. For this, furrows are opened 5 cm deep on the one side of the ridge and then fertilizer mixture is applied and covered with soil. The basal application is repeated once in a year for sustained higher yields.

b) **Top dressing**: 30 days after planting and after each cut, 75 kg N/ha is applied.

Seeds and Sowing

a) Seed rate: 33,000 rooted slips or stem cuttings/ha

b) Spacing: 60 x 50 cm.

c) Sowing: The field is irrigated through furrows and one rooted slip or stem cutting is planted per hole at a depth of 3-5 cm on one side of the ridge.

After cultivation

Hand weeding or hoeing and weeding on 30th day is done. Then the gaps are filled to maintain the population. Subsequent weeding may be carried out preferably after each harvest. Earthing up once after three cuts and removal of dried tillers and quartering once in a year is recommended.

Irrigation

Life irrigation is given on 3rd day and thereafter once in 10 days or as required. Top dressing after each harvest is done followed by irrigation.

Harvest

First cut is carried out 60-75 days after planting and subsequent cuts once in 45 days. In case of sewage or high N containing effluent irrigation, the harvest intervals may be increased to 55-60 days to minimize the nitrate and/or oxalate problems.

Green fodder yield

NB-Hybrid Co (CN) 4 and CO (BN)5: 350-400 t /ha / yr. in 7 harvests

Questions

I. Choose the best from the choices given

1. The recommended spacing for Napier bajra hybrid is

   a) 50 × 15 cm b) 30 × 10 cm

   c) 45 × 15 cm d) 60 × 50 cm

2. Bajra napier hybrids can be planted in

   a) June-July b) Aug - Sept

   c) Throughout the year d) Dec - Jan

3. The seed rate for Bajra napier hybrid grass is

   a) 23000 cuttings / ha b) 33000 cuttings / ha

   c) 43000 cuttings / ha d) 53000 cuttings / ha

4. In case of sewage or high N containing effluent irrigation, the harvest intervals may be increased to 55-60 days to minimize

   a) Phytase b) Mimosine

   c) Nitrate d) HCN

5. For Bajra napier hybrid, the following quantity of FYM should be applied as basal

   a) 10 t / ha b) 15 t / ha

   c) 20 t / ha d) 25 t / ha

II. Fill in the blanks

1. The fertilizer recommendation for Bajra Napier hybrid is ________ kg NPK / ha.
2. ___________ kg of Nitrogen should be applied to Bajra Napier hybrid after each harvest.
3. In bajra napier hybrid grass, first cut is done at __________DAP.
4. In bajra napier hybrid grass subsequent cuttings after first cut can be done once in __________.
5. The regeneration capacity is quick in _________ grass.

III. Write short notes

1. Nutrient management for Co (BN) 5
2. Weed management in Bajra Napier hybrid grass
3. Harvesting in Bajra Napier hybrid grass
4. Special features of Co (BN) 5
5. Varieties of Bajra Napier hybrid grass

17

Fodder Cowpea, Cluster Bean - Origin, Geographic Distribution Economic Importance, Soil and Climatic Requirement, Varieties Cultural Practices and Yield and Fodder Preservation

COWPEA (*Vigna unguiculata* L.)

Cowpea is a quick growing leguminous forage crop. It is usually grown mixed with cereal fodders and grasses to improve the nutritive value of the herbage. It contains 20 – 24 % crude protein, 43 - 49 % neutral detergent fibre, 34 – 37 % acid detergent fibre, 23– 25 % cellulose and 5 – 6 % hemicelluloses on dry matter basis. The digestibility of cowpea fodder is above 70%. Cowpea can be grown under partial shaded conditions. It is an excellent cover crop, which suppresses weeds and enriches the soil. Cowpea requires warm climate with good atmospheric humidity. It can be grown in *kharif* as well as in *zaid* season.

Origin

Vavilov considered India as the center of origin of cowpea and Africa and China as secondary centers. But historical and archeological evidences strongly suggested that Africa is most probably the place where the present day cultivated cowpea evolved.

Distribution

It is most widely raised in the tropics and subtropics, particularly in Africa, the Indian subcontinent, southeast Asia and tropical America. Cowpea is mainly grown in Africa. About 90% of the world area is in Africa. In India, it is mainly cultivated in Central and peninsular regions. In northern India it is grown in Uttar Pradesh, Punjab, Delhi and Haryana.

Soil

Cowpea can be grown on variety of soils. The plants prefer light soils. Loam and sandy loam soils with good drainage are most suitable for good crop growth. Field should be prepared by two cross harrowing and planking so as to get a leveled and weed free seed bed for quick germination and faster initial growth.

Varieties

Varieties	Areas of cultivation	Green fodder (t/ha)
EC-4216	North zone	30-35
UPC-5286	Whole country	35-40
IFC-8503, EC-4216	North, West and Central India	30-40
UPC-5287	North zone	30-45
CO 5, CO(FC) 8	Tamil Nadu	30-45
UPC-4200	North-East region	30-35
GFC-1, GFC-2, GFC-3, GFC-4	Gujarat	25-35
Sheweta	Maharashtra	35-40
Bundel lobia-1	Whole country	30-35
Bundel lobia-2	North west zone	35-40
UPC-618, UPC-622	North- west, North-east and Hill zone	35-45

Sowing time

In irrigated areas, sowing can be done during summer while in rainfed areas, it can be done after commencement of rains. Its sowing extends from March to middle of July. In southern region, sowing for fodder may be done throughout the year.

Seed rate and sowing method

A seed rate of 35-40 kg/ha is sufficient for its proper plant population. The sowing should be done in lines at an inter row spacing of 25-30 cm. The seed should be sown with seed drill or behind the plough at a depth of 2-3 cm.

Cropping systems

Promising cropping systems	Green fodder production potential (t/ha)
Central and western region	
Cowpea – Sorghum + Cowpea – Berseem (semi-arid sandy soil)	211
Sorghum + Cowpea – Berseem + Mustard - Sorghum + Cowpea (sub humid, black soil)	168

Eastern region	
Pearl millet + Cow pea – Maize + Cowpea –Oats	102
(Sub humid, red acidic soils)	
Maize + Cowpea –Sorghum + Cowpea –Berseem + Mustard	96
(sub humid, red acidic soils)	
Southern region	
Sorghum + Cowpea – Maize + Cowpea – Maize + Cowpea	110

Nutrient management

Cowpea is a leguminous crop and has capacity to fix atmospheric nitrogen. However, 20 kg N and 60 kg P2O5 /ha should be applied at the time of sowing for good crop growth. In sulphur deficient soils (below 10 ppm), 20-40 kg sulphur per hectare is recommended for quality fodder biomass production.

Water management

Normally the *Kharif* season crop does not require irrigation except in case of long dry spells in which the crop should be irrigated at an interval of 10-12 days. But summer crop requires 6-7 irrigations at 8-10 days interval.

Weed management

In general, *Kharif* crops are densely infested with weeds due to conducive situation for growth. In cowpea, the weed problem is severe in early stages. After 30 days the crop covers the land area and thus, problem of weed infestation is minimized. One manual weeding or hoeing with weeder cum mulcher at 3 weeks crop stage is effective to check weed growth. Pre-plant soil incorporation of Trifluralin or Fluchloralin @ 0.75 kg a.i. /ha has been found to be a useful chemical weed management method to arrest weed growth.

Harvest

Rainy season crop is harvested after 50-60 days of sowing at 50% flowering stage whereas summer crop requires few more days and should be harvested after 70-75 days of sowing.

Yield

Under irrigated condition, cowpea crop yields 25-30 t/ha green biomass. But under rainfed condition, 15-20 t/ha a green fodder can be obtained.

CLUSTER BEAN (*Cyamopsis tetragonaloba* L. Taub)

Cluster bean is a leguminous annual and erect herb growing to a height of 1-2 m. It is cultivated in all parts of India. It is a drought resistant crop and suitable for cultivation in arid and semi-arid regions. Being a legume, it produces good palatable green fodder with 17-20 % crude protein, 42-48 % neutral detergent fibre, 37-42 % acid detergent fibre, 23.5-25.3 % cellulose and 8-12 % hemi cellulose content.

Origin

The origin of cluster bean is Tropical Africa. According to Vavilov, clusterbean is native of India

Soil

It is a soil restorative crop, especially suitable for light sandy or alluvial soil. The facility of adequate drainage is ideally suited for its growth. The field should be prepared by 2-3 harrowing to ensure a leveled and weed free soil surface.

Climate

It is a tropical drought resistant crop. It is grown in drier regions receiving 300-400 mm of rainfall. The ideal temperature for this crop is 25-32^0 C.

Season

Under irrigated conditions, *summer* crop should be sown in March-April and rainy season sowings should be done in June-July with the onset of monsoon in north Indian conditions. However, in southern parts of the country winter sowings can be done in October-November.

Varieties

Varieties	Areas of cultivation	Green fodder (t/ha)
FS-277	Entire guar grown tract	30-35
HFG-110, HFG-156	Entire guar grown tract	25-30
Guara-80	North-West zone	20-30
Maru guar	Western Rajasthan	25-30
Bundel Guar-1, Bundel Guar - 2, Bundel Guar-3	Entire guar growing tract	30-35

Seed rate and sowing

Sowing is done in 25 cm apart lines using a seed rate of 30-35 kg/ha. In dryland semi-arid regions, where moisture stress is common, sowing is recommended in 30 cm apart rows with reduced seed rate of 25-30 kg/ha to have a low plant population. The seed should be sown with *Kaira* under irrigated conditions and *Pora* in dry lands.

Cropping system

Some of the cropping systems of Cluster bean are Sorghum + Cluster bean – Oat, Pearl millet + Cluster bean – Oat and sole cluster bean in newly developed sites.

Nutrient management

It is a leguminous crop and has ability to fix atmospheric nitrogen. Therefore, only 20 kg N + 50 kg P_2O /ha at the time of sowing should be applied.

Water management

Summer crop requires 3-4 irrigations, whereas, rainy season crop generally does not require any irrigation. However, if long dry spell prevails, one or two irrigations may be provided. *Rabi* season crop in southern India needs 3-4 irrigations.

Weed management

Pre-plant soil incorporation of Nitralin @ 0.75 kg a.i. /ha has been found effective. One operation with weeder cum mulcher at 3-4 weeks crop stage is very useful for checking weed growth.

Harvest

The crop should be harvested at bloom to pod formation stage (60-75 days after sowing).

Yield

A good crop of cluster bean yield 25-30 tonnes green fodder /ha.

FODDER PRESERVATION

Fodder preservation is conservation of surplus fodder produced during the regular seasons in any form without affecting the nutrient content and palatability much for use in the lean seasons.

Necessity to preserve fodder

Fodder production is seasonal. But, demand for fodder is uniform throughout the year. There is a glut of fodder during peak seasons of growth and to regulate the fodder supply during the lean seasons, fodder has to be preserved. In drought prone areas where crop failures are frequent, fodder preservation is the only way to tackle fodder scarcity under such situations. Milch animals particularly cattle require green fodder throughout the year. To supply the same, green fodder should be preserved in a succulent state.

There are two important methods of fodder preservation, namely Silage and Hay

Each method has its own limitations and advantages but ensiling is preferred on the basis of fodder quality.

Silage

Silage is the preserved green fodder in its succulent state by packing fresh fodder and allowing it to ferment under anaerobic conditions, without undergoing much loss of nutrients.

The process of making silage is called Ensiling.

Hay

It is defined as the conversion of green forage into dry form without affecting the quality of the original material, which can be safely stored for fairly long periods.

Silage making

Silage is defined as the green succulent roughage preserved under controlled anaerobic fermentation in the absence of oxygen by compacting green chops in air and watertight receptacles. The ensiling process, in the absence of oxygen, leads to the fermentation of water-soluble carbohydrates to produce organic acids, which increases acidity of the materials.

Chaffing

Ensiling chaffed green fodder gives quality silage as it facilitates easier and compact packing and as a result the air pockets are reduced to a minimum. The forage material for ensiling has to be chaffed in a chaff cutter into pieces of 1-3 cm size. Salt and jaggery / molasses, each at the rate of 1% on green weight basis is used as preservatives. A saturated solution has to be prepared by dissolving in limited quantity of water and sprinkled on the chaffed material and mixed well.

Filling the silo

The prepared forage material should be filled layer by layer, 15-20 cm, thickness at a time, in the silo and well pressed by manual trampling to remove air pockets. Filling must be completed quickly and the silo should be closed tightly by polythene sheet and sealed using mud as tightly as possible.

Opening the silo

Silo pit can be opened for feeding 30-35 days after ensiling. It can be easily stored for 3-6 months or even more if anaerobiosis is maintained. Once the silo is opened then it has to be used as quickly as possible to avoid spoilage. To reduce spoilage due to aerobiosis, after opening the silo and removing the required quantity of silage, the top must be covered with the excess gunny and polythene and a small weight (brick) may be placed to reduce air entry and to retain compaction.

Silage feeding

Silage is normally fed during lean periods when green fodder availability is scarce. Usually, one fourth of the green fodder ration is supplemented with silage and other roughage meets the rest. For example, an animal fed with 35 kg green fodder per day, 9 kg silage is sufficient and other roughage ad lib is recommended for better health and milk production. Feeding excessive silage or only silage may upset the stomach (rumen) conditions and cause ill effects. It is advisable to feed 6-9 kg silage depending on the animal weight and milk yield.

Points for consideration while ensiling

1. The dry matter content of the material to be ensiled must be between 30 and 45%.
2. More succulent materials may be allowed to wilt in the field itself before ensiling.
3. Polythene layering on all sides of the pit improves quality.
4. Filling should be done on a clear day and as quickly as possible.
5. Filling should be done in layers of 20-30 cm thickness at a time and as uniformly as possible.

6. Compaction must be as perfected as possible. Here trampling may be more useful to remove air pockets and to facilitate even packing of the material.
7. Top must be convex or dome shaped, covered with dry straw and sealed with mud plaster or covered with alkathene or polythene sheets.
8. Under ideal conditions it can be stored easily for 1 year.

Characteristics of good silage

1. There should not be any mould growth.
2. It should be golden brown / greenish yellow in colour.
3. It should have pleasant fruity odour / an acceptable aroma.
4. Liked very much by animals because of mild-acidic taste and pleasant aroma.
5. It should be free flowing and non-sticky texture.
6. Silage increases the palatability by 3-4%.
7. It should have increased nutritive value
8. The silage pH should be around 4.0-4.5
9. Lactic acid is proportionately more than the other acids (3 Butyric acid is very low and is in the range of 0.2-0.5%)
10. In general nitrate-N decreases and ammoniacal – N increases upon ensiling.
11. Ammoniacal N should not exceed 9-15% of the total – N.

Advantage of silage

1. Comparatively less loss of nutritive value of the fodder.
2. It is more suited in such of those seasons when weather is not conducive for haymaking.
3. Thick stemmed crops like sorghum and maize are better utilized
4. Comparatively highly palatable and nutritious
5. Organic acids produced during ensiling are similar to those organic acids produced in the digestive tract of the ruminants

Hay making

Hay refers to cereals, grasses or legumes that are harvested at the appropriate stage, dried and stored. High quality hay is light or grayish in colour, leafy, pliable and free from mustiness. Hay making is the most common and easy method of preserving seasonal excess of green fodder and the only method of preserving farm by –products. The principle involved in hay making is to reduce the water content of the herbage so that it can safely be stored in mass without undergoing fermentation or becoming mouldy. Legume hay, non-legume hay and mixed bay are the major three types of common hays used in livestock feeding.

Questions

I. Choose the best from the choices given

1. The recommended row spacing for cowpea is

   a) 50 cm b) 30 cm

   c) 45 cm d) 60 cm

2. The green fodder yield of cowpea under irrigated condition is

   a) 10-15 t / ha b) 20-25 t / ha

   c) 25-30 t / ha d) 35-40-15 t / ha

3. The seed rate for fodder cowpea is

   a) 35-40 kg / ha b) 20-25 kg / ha

   c) 10-15 kg / ha d) 50-60 kg / ha

4. The green fodder yield of clusterbean under rainfed condition is

   a) 25-35 t/ha b) 35-45 t/ha

   c) 10-15 t/ha d) 5-10 t/ha

5. Under ideal condition, silage can be stored for

   a) 1 yr. b) 2 yrs.

   c) 3 yrs. d) 4 yrs.

II. Fill in the blanks

1. The fertilizer recommended for fodder cowpea is ________ kg NPK / ha.
2. ___________ kg P_2O_5 should be applied to clusterbean at the time of sowing.
3. The seed rate for clusterbean for fodder purpose is _________ kg / ha.
4. The maximum quantity of silage that can be fed to milch animal is ___kg / day
5. The ammoniacal N should not exceed _________% of the total N in silage

III. Write short notes

1. Nutrient management for Cowpea
2. Weed management in Clusterbean
3. Weed management in Cowpea
4. Qualities of good silage
5. Steps in silage making